T0332479

AGROFORESTRY EDUCATION AND TRAINING

FORESTRY SCIENCES

Volume 35

Agroforestry Education and Training: Present and Future

Proceedings of the International Workshop on Professional Education and Training in Agroforestry, held at the University of Florida, Gainesville, Florida, USA on 5–8 December 1988

Edited by

P.K.R. NAIR, H.L. GHOLZ and M.L. DURYEA

Department of Forestry
Institute of Food and Agricultural Sciences
University of Florida
Gainesville, Florida, USA

Reprinted from Agroforestry Systems 12(1)

Kluwer Academic Publishers
DORDRECHT / BOSTON / LONDON

ISBN 0-7923-0864-6

Published by Kluwer Academic Publishers,
P.O. Box 17, 3300 AA Dordrecht, The Netherlands.

Kluwer Academic Publishers incorporates
the publishing programmes of
D. Reidel, Martinus Nijhoff, Dr W. Junk and MTP Press.

Sold and distributed in the U.S.A. and Canada
by Kluwer Academic Publishers,
101 Philip Drive, Norwell, MA 02061, U.S.A.

In all other countries, sold and distributed by Kluwer Academic Publishers Group,
P.O. Box 322, 3300 AH Dordrecht, The Netherlands.

Technical editor: Nancy Dohn

Printed on acid-free paper

Printed in the Netherlands

Contents

Preface 3

Acknowledgements 5

Inaugural address

E.T. York, Jr.: The importance of agroforestry education and training 7

Keynote papers

E. Zulberti: Agroforestry education and training programs: an overview 13

H.A. Steppler: Agroforestry education 41

R.F. Fisher: Agroforestry training: Global trends and needs 49

K.G. MacDicken and C.B. Lantican: Resource development for professional education and training in agroforestry 57

Regional/Country reports

E.O. Asare: Agroforestry education and training: an African experience 71

R. Sands: Education in agroforestry at the University of Melbourne 81

Wang Shiji: A brief account of professional education and training in agroforestry in China 87

H.-J. von Maydell: Agroforestry education and training in European institutions 91

Kirti Singh: Status of agroforestry education in India 97

A. Sumitro: The status of professional agroforestry education and training in Indonesia 103

J.C.L. Dubois: Agroforestry education and training in Latin America 107

S.T. Warren and W.R. Bentley: Expanding opportunities for agroforestry education in the U.S. and Canadian Universities 115

Workshop summary and synthesis

J.P. Lassoie: Towards a comprehensive education and training program in agroforestry 121

Workshop recommendations 133

List of participants 141

Preface

Interest and initiatives in agroforestry education and training, as in other aspects of agroforestry development, have increased tremendously during the past decade. Coordination of such educational activities was initiated by the first international workshop on education in agroforestry organized by the International Council for Research in Agroforestry (ICRAF) in December 1982, at Nairobi, Kenya. Since then, agroforestry has been incorporated into the curricula of many educational and training institutions around the world. Moreover, several institutions have developed entire academic programs specifically in agroforestry. However, most of these activities are still isolated initiatives, without common strategies or philosophies. This second international agroforestry workshop was therefore planned to provide a forum for reviewing progress, sharing programs and experiences, and planning and coordinating future directions in agroforestry education and training. The main objectives were to review the on-going programs, to assess the scope of professional education and training in relation to the perceived needs of trained personnel, to recommend guidelines for further program development, and to establish networking among institutions and agencies involved in agroforestry education and training.

These proceedings contain the keynote papers, regional/country presentations and conclusions and recommendations of the International Workshop on Education and Training in Agroforestry held at the University of Florida, Gainesville, USA in December, 1988. The strategy of the workshop was to facilitate focused discussion on identified issues by an invited group of world leaders in agroforestry education and training. First, four keynote papers presented global reviews of the (1) current status and trends in agroforestry education and training, (2) current and future needs in agroforestry education, (3) current and future needs in training, and (4) current and future needs in resource development and networking. Following these global overviews eight regional/country presentations reported the present status of education and training at specific institutions throughout the world. Key issues and questions identified in these sessions were addressed in meetings of smaller working groups on education, training and resource development. These intensive sessions produced recommendations which were critiqued at the final plenary session. The final workshop recommendations are a group effort and product of the insightful thoughts and discussions during the workshop.

We hope that this Proceedings and its recommendations provide a valuable guide for further development and continued growth of agroforestry education and training in the world.

University of Florida
Gainesville, Florida
January 1990

P.K.R. Nair
H.L. Gholz
M.L. Duryea

Acknowledgements

Many individuals and organizations contributed to the success of this workshop. The attendance by a distinguished gathering of world leaders in agroforestry education, many relying on their own financial support, was a testament to the importance of the event. We thank all these invited participants for the energy, diligence and diverse skills that they brought to the workshop, truly making it a success.

The International Council for Research in Agroforestry (ICRAF) was an official cosponsor; in addition to the participation of two of its senior staff, it also sponsored four other participants from Africa. A grant from the Ford Foundation office in India enabled two participants from India to attend. The Ford Foundation office in Brazil supported the attendance of four delegates from that country. The Winrock International Institute sent two of its officials to the workshop – one from Arkansas, USA, and the other from Bangkok, Thailand – and sponsored the travel of the participant from Malaysia. The Forestry Support Program of the U.S. Department of Agriculture (USDA-FSP) was represented by two officials from Washington, D.C. and sponsored the participants from Costa Rica and Peru. The Center for International Programs, Institute of Food and Agricultural Sciences (IFAS), University of Florida, supported six participants and The Center for Latin American Studies, University of Florida, supported a delegate from the Dominican Republic. Sincere thanks are due to all these organizations and institutions for their support.

Special thanks are extended to the Workshop Organizing Committee, which included Dr. J. Gordon, Yale University, Dr. D.V. Johnson, USDA Forestry Support Program, Dr. E. Zulberti, ICRAF, and Drs. P. Hildebrand, C.P.P. Reid, and M. Swisher, University of Florida, for their guidance during the planning phases of the workshop.

Administrative and logistical arrangements for the workshop were handled by the Office of Conferences and Institutes, IFAS, University of Florida.

We are grateful for the production assistance of Nancy Dohn for technical editing of the proceedings and to Elizabeth Gaylord for typing the manuscript.

Finally, we thank the Editor-in-Chief and Kluwer Academic Publishers for publishing the workshop proceedings as a special issue of *Agroforestry Systems*.

Agroforestry Systems **12**: 7–12, 1990.

Inaugural address

The importance of agroforestry education and training

E.T. YORK, JR.
University of Florida, Gainesville, FL 32611, USA

Abstract. Sustainable agriculture should involve the successful management of resources for agriculture to satisfy changing human needs, while maintaining or enhancing the quality of the environment and conserving natural resources. The rapid depletion of forest resources and agricultural sustainability – two major global concerns – could be impacted significantly through agroforestry practices. Widespread application of agroforestry concepts and techniques offers great potential for helping to alleviate critical shortages of fuelwood and contributing to sustainable farming systems. There is a vital need to broaden the knowledge base of the subject of agroforestry in order to provide a more substantive basis for effective teaching and training programs. In a typical university organization, it is fairly easy to see how interdisciplinary research teams can be brought together to work on agroforestry projects. But how do we structure the education and training programs? The purpose of this conference is to address these issues and to guide the further evolution and development of agroforestry.

Introduction

The topic of this conference is of enormous importance – with great potential implications for our planet and its inhabitants. Indeed, it is a concept, the time for which has truly come.

There are two major problems of global concern which could be impacted significantly through agroforestry practices. The first one is the rapid depletion of forest resources – especially those used for fuel – and the other, agricultural sustainability.

The depletion of forest resources

It is estimated that 11 million ha of tropical forest are destroyed annually as a result of clearing land for fuelwood and for agricultural operations. An additional 4.5 million ha or more are estimated to be cleared through commercial timber operations annually [Repetto 1988].

An FAO study indicates that tropical trees are being cut much more rapidly than reforestation or natural processes may replenish them. Indeed, this study suggests that for every ten hectares cleared, only one is replaced [Lanly 1982].

More than half of the world's tropical forests have disappeared since the turn of the century. Today more than half of the developing world's population lives in 56 of the most critically affected countries.

I will not discuss all of the consequences of such deforestation, including the loss of unique ecosystems which may directly contribute to the extinction of plant and animal genetic resources. Neither will I attempt to address the potentially serious environmental implications of such deforestation, including the CO_2 related 'greenhouse effect'. Rather I want to explore some other consequences which have direct implications to the subject at hand, specifically the emerging fuelwood crisis throughout the developing world.

A few statistics may be helpful to illustrate the magnitude of this problem. It is estimated that more than two-thirds of the people throughout the developing world rely on wood for cooking and heating. Indeed, in many developing countries, wood provides more than two-thirds of the energy used for all purposes. An estimated 250 million people in developing countries live in areas of fuelwood shortage. Recent FAO estimates suggest that by the year 2000 there will be almost 2.5 billion people living in the developing world under acute scarcity or deficit fuelwood conditions [FAO 1983]. Sixty percent of those will be in Asia, not including China.

The consequences of such shortages are serious and varied. For example, such shortages place additional demands on other energy sources, which are also limited; there are also many social and health-related implications.

It is estimated that in some parts of the Himalayas and the African Sahel, women and children spend between one hundred and three hundred days a year gathering fuelwood. A recent study in Nepal by the International Food Policy Research Institute has indicated that the destruction of forest resources has added one hour per day to the time required for women to collect firewood. As women put more time into gathering firewood, they have less time for working in the fields. Shortages also contribute to a decline in the nutritional status of children through the reduced time of the mother for cooking and food preparation. It often becomes necessary to substitute quick-cooking cereals for more nutritious, but slower-cooking foods such as bean. Health problems are increased because time for boiling water becomes a luxury.

Urbanization further exacerbates the fuelwood shortage since people in cities rely more on charcoal than wood, and since half the energy in wood is lost while it is converted to charcoal by burning. The impact of this urbanization trend is reflected by some recent estimates by the World Bank which indicate that by the year 2000, 50 to 70% of West Africa's fuelwood consumption will be in urban areas.

Widespread application of agroforestry concepts and techniques offers great potential for helping to alleviate some of the critical shortages of fuelwood throughout the developing world.

Agricultural sustainability

There is increasing interest in the subject of sustainable agriculture. An international conference was held on this subject at The Ohio State University, USA in September 1988. The World Commission on Environment and Development (the so-called Brundtland Commission) addressed this subject at some length in its recent report, titled, 'Our Common Future' [WCED 1987]. For the past two years, I have chaired a panel of the Technical Advisory Committee of the Consultive Group on International Agricultural Research (TAC, CGIAR) on agricultural sustainability. This effort led to the development of a comprehensive report on the subject which was presented to the CGIAR in Berlin May 1988.

Since this is a relatively new area of concern, and since agroforestry is so relevant to consideration of sustainability issues, let's address this subject briefly.

First, a quote from the report of the World Commission on Environment and Development puts the problem in perspective: 'The next few decades present a greater challenge to the world's food systems than they may ever face again. The effort needed to increase production in pace with unprecedented increase in demand, while retaining the essential ecological integrity of food systems, is colossal, both in its magnitude and complexity. Given the obstacles to be overcome, most of them man-made, it can fail more easily than it can succeed.'

The increasing global interest in issues related to agricultural sustainability grows out of concerns that many factors are contributing to the degradation of the environment. Intensification of traditional agricultural systems to meet increasing needs often has undesirable environmental or ecological consequences, such as soil erosion, salinity, waterlogging and the contamination of aquifers with chemicals.

Some of these factors prompted the TAC to recommend in 1987 that 'sustainable' be included in CGIAR's basic goal of increasing food production, and that greater emphasis be placed on sustainable production in the future work of the International Centers. The CGIAR, in turn, asked the TAC to explain what it meant by 'sustainability', which led to the development of the report referred to above.

There have been many definitions of sustainable agriculture. TAC proposed a concept which has been accepted by the CGIAR, by USAID and others. The TAC suggested that the goal of sustainable agriculture be to maintain production at levels necessary to meet the increasing aspirations of an expanding world population without degrading the environment. Sustainable agriculture should, therefore, involve the successful management of resources for agriculture to satisfy changing human needs, while maintaining or enhancing the quality of the environment and conserving natural resources.

In this context, what can agroforestry contribute? Contributions, both direct and indirect, can be many and varied.

The use of agroforestry techniques to accommodate fuelwood needs may make some indirect contributions to the improvement of soil fertility. For example, when fuelwood is in short supply, rural people are forced to use plant residues and animal manures for fuel, rather than returning them to the soil. This contributes to mining the soil of its native supply of nutrients, especially under conditions where fertilizer use is limited. FAO estimates that in Africa ten times as many nutrients are being removed in crops as are being put back in the soil through the use of organic and mineral fertilizers. Such serious mining of soil nutrients will make the achievement of sustainable production impossible unless the nutrients are replenished through external inputs.

In systems including trees and field crops, trees continuously contribute to soil organic matter through shedding of their leaves and roots. This organic matter improves soil structure, fertility and waterholding capacity. Deep-rooted trees absorb nutrients from greater depths and deposit them on the surface in organic matter so that they are more available to shallow-rooted crops.

Trees used in windbreaks can also increase water availability by reducing windspeed and thereby lowering evapotranspiration. Tree canopies lower the impact of heavy rainfall, reducing runoff and increasing water infiltration into the soil. Moreover, the shade provided by trees lowers soil temperatures, reducing evaporation and slowing the decomposition of organic matter. Leguminous trees may provide still further advantages by adding nitrogen to the soil through plant residues.

Many traditional agricultural production systems were sustainable for centuries in terms of their ability to maintain relatively stable levels of production. However, the needs and increasing aspirations of expanding numbers of people have forced changes in land use and imposed excessive demands on natural resources, making modified farming systems unsustainable. Increasing efforts must be devoted to finding ways in which the sustainability of more productive agricultural systems might be increased. To put that in the context of this workshop, new and improved agroforestry systems must be devised to meet these needs. Moreover, the productivity of existing or traditional systems must also be improved.

While the productivity of traditional farming systems may become inadequate to meet increasing needs, the principles on which they are based have permitted them to persist for generations. Research has already led to considerable understanding of the strengths and limitations of many aspects of traditional systems. Others are more difficult to analyze and research methodologies for this purpose are still evolving. There are dangers both in disregarding the principles of traditional systems and in assuming that, because they are appropriate in some circumstances, they will be appropriate in all others.

Generally, there is a great need for more research on the crops and livestock in production systems that exploit the principles of agroforestry in order to optimize production and insure sustainability. There is also need to investigate more fully the wide range of multipurpose trees and shrubs that might find a place in production systems of this type.

Agroforestry education and training

The focus of this workshop is on agroforestry education and training. Realize however, that there is a vital need to broaden the knowledge base of the subject in order to provide a more substantive basis for the development of effective teaching and training programs.

As indicated earlier, much can be learned from traditional systems. But more research is essential in determining how traditional systems can be improved and made more productive. We also need research that might lead to new systems that could be even better than many of the traditional systems. The various combinations of woody species and crops which could constitute such systems are almost unlimited.

I recognize and applaud the agroforestry research currently underway, at the International Council for Research in Agroforestry (ICRAF), for example, and at some of the other International Centers such as the International Institute of Tropical Agriculture (IITA), Nigeria. Work at IITA on alley farming would appear to offer tremendous potential. But it is disappointing that farmers have not adopted such practices more readily. It is very important to know why. Obviously more research is needed in the social sciences, to help address these issues.

In a typical university organization, it is fairly easy to see how interdisciplinary research teams could be brought together to work on agroforestry projects. But what about education and training programs?

Where, within a university structure, do such programs logically belong? Are we far enough along to begin to think about departments of agroforestry? If so, what would be the makeup of such departments and where should they be located? Or, ignoring the question of departmental status, if an institution wanted to establish a program in agroforestry, where should that be located?

Taking our own institution (the University of Florida's Institute of Food and Agricultural Sciences – IFAS) as an example, should agroforestry be in the School of Forest Resources and Conservation or in the College of Agriculture? If in agriculture, where should it be centered? How do you encourage personnel from various departments within the college to participate? How do you involve social scientists located outside the college?

There are many issues to pursue in a conference such as this. We look forward to the product which should help guide the further evolution and development of agroforestry.

References

FAO (1983) Fuelwood Supplies in Developing Countries. FAO, Rome

Lanly JP (1982) Tropical Forest Resources. Forestry Paper No. 30. FAO, Rome

Repetto R (1988) The Forest for the Trees? Government policies and the misuse of forest resources. World Resources Institute, Washington, DC

WCED (1987) Our Common Future: Report of The World Commission for Environment and Development. Oxford University Press, London

Agroforestry Systems **12**: 13–40, 1990.

Keynote paper

Agroforestry education and training programs: an overview

ESTER ZULBERTI
International Council for Research in Agroforestry (ICRAF), P.O. Box 30677, Nairobi, Kenya

Abstract. Agroforestry has been incorporated in education and training programs at an unprecedented level since 1982. A survey of educational institutions conducted by ICRAF in 1987 revealed that agroforestry is found as an option for specialization in undergraduate as well as in postgraduate M.Sc. diploma programs in forestry, agriculture, natural resources, and others. Courses and special seminars in agroforestry are organized in degree programs. Full undergraduate and postgraduate programs in agroforestry are being formulated and implementation started in quite a few universities, and many students are choosing agroforestry-oriented research projects for their dissertations. A good setting for higher degree training in agroforestry requires, however, staffing from combined faculties of at least agriculture, animal science and forestry; faculty commitment to a farming systems approach; and inter-departmental cooperation in teaching and research. It was difficult to assess whether these and other elements are present in existing programs where agroforestry has been incorporated. Emerging trends indicate that traditional forestry programs are broadening the scope of the discipline (from forests to integrated land-use systems) while agriculturists are recognizing that trees play important roles as soil improvers and protectors, fodder, food, fuel and other domestic and commercial purposes. New institutional structures are evolving to allow for educational programs with coursework and research projects spanning many disciplines. Nondegree training in agroforestry has seen an upsurge of activities equal, if not larger, to that in education. Attempts are being made by different institutions worldwide to inventory training opportunities; still the collection and dissemination of information is difficult. Efforts are needed at the international, regional, and national levels, to address training issues that if addressed collectively can improve the quality and effectiveness of human resource development efforts. ICRAF's approach to promote agroforestry research through education and training is an example of an action program currently under application.

Introduction

The history of agroforestry as a science is short, 15 years at the most. However, agroforestry is only a new term for age-old practices of integrated land-use in almost all parts of the world. Even though still referred to as a 'new area' of scientific activity, the number of national and international research institutions, development and donor agencies, non-governmental agencies (NGOs) and others that have taken up agroforestry or agroforestry-related activities has increased rapidly over the last few years.

A group of distinguished research scientists and agroforestry leaders worldwide met in Nairobi in July 1987, on the occasion of the Tenth Anniversary of the International Council for Research in Agroforestry (ICRAF). Their evaluation of the recent history [Steppler and Nair 1987] confirmed that agroforestry has come of age and it is here to stay.

Land-use circumstances under which existing agroforestry technologies are applied and the interactions between system components are not, in most cases, properly understood. Research results in agroforestry are, thus, scarce if compared to those available for cash and food crops.

The scarcity of knowledge on existing and potential agroforestry technologies in most tropical and subtropical countries has been largely attributed to the lack of institutional capability to confront pressing land-use problems requiring an innovative approach. Existing rigid structures in compartmentalized, conventional, disciplinary-oriented institutions in research and development programs, have been identified as major limitations for the development of integrated land-use activities. Insufficient resource allocation compounds the problem [Lundgren 1987]. It remains a fact that in spite of efforts made by international centers to integrate forestry research more closely with agriculture, livestock, social sciences, and other disciplines, few research institutions have achieved a similar integration at the national level. Most agricultural research programs continue to focus on rice, wheat and maize breeding, while forestry research continues to concentrate on tree improvement or improved utilization [Schuh 1987].

The research constraint is further aggravated by a shortage of trained and experienced professionals with knowledge and skills to integrate several disciplines that together must be combined in researching, planning and managing agroforestry [Contant 1980, Huxley 1980].

The development and implementation of agroforestry education and training programs have been impaired by several factors, among which the following have often been mentioned: rigid institutional structures that, as is the case in research and development, have not allowed for programs to cut across various disciplines; insufficient information on verified scientific methods; lack of appropriate instructional materials; and, lack of adequately prepared lecturers and trainers [Zulberti 1987].

To summarize, agroforestry research and development is constrained by rigid institutional structures and lack of trained professionals. At the same time, the training of professionals in agroforestry is equally constrained by insufficient research methodology and rigid institutional structures. Yet, agroforestry is fast becoming incorporated in education and training programs as an experimental science that can be taught. The paradox presents, no doubt, an interesting challenge to this workshop.

The critical issue is not so much the identification of recipes to teach agroforestry worldwide – as programs will have to respond to different human-resource development needs and country conditions – but to understand the rationale behind approaches developed by institutions that have taken the initiative to incorporate agroforestry in existing education and training programs. The task of the workshop participants is, therefore, to share those experiences, positive or otherwise, that may help identify paths and cost-effective ways to better respond to the human-resource development needs of developing and developed countries in the future.

Fortunately, we are not starting from zero. An international workshop on 'Professional Education in Agroforestry' was organized by ICRAF and the German Foundation for International Development (DSE) in Nairobi in 1982 to debate and establish priorities, plans and procedures for action, mainly in agroforestry professional education but with obvious implications for training programs as well. That was the first attempt to bring together experts representing the fields and disciplines related to agroforestry education on a worldwide basis. A rich source of information, although still far from complete, was uncovered then about where and how agroforestry education was being done [Zulberti 1987].

In this paper, an attempt is made first to review progress made in the incorporation of agroforestry in education and training programs since the Nairobi Workshop in 1982 up to 1988. The review is based on information currently available to ICRAF. A distinction is made between 'education' and 'training'. Education refers to broader, longer-term studies to achieve a higher academic qualification. Training refers to relatively short term, in-service study to achieve a higher level of technical knowledge and/or skills.

ICRAF's approach to promote agroforestry research through education and training is also presented as an example of an action program currently under application. Finally, some general ideas on directions are mentioned that may produce some food for thought.

From the past to the present

The past

It is deemed appropriate at this time to extract the main conclusions and recommendations of the Nairobi workshop as an *aide memoire* and as a starting point to the present workshop.

In assessing the overall agroforestry education situation as it was in 1982, the participants at the workshop felt that there was a great deal of enthusiasm to include agroforestry in existing programs. Full programs in agroforestry were, however, only offered in nondegree training programs. At the undergraduate and postgraduate levels, agroforestry was mostly included as an element of existing courses in forestry programs. Research in agroforestry was scarce.

Among the major weaknesses in teaching agroforestry, the following were mentioned: the definition of agroforestry varies in different parts of the world; there is inadequate knowledge about agroforestry as a land-use system (e.g., solid data on land tenure problems, farmer needs and traditional systems, etc.); rigid institutional structures make the cutting-across discussions difficult; and, there is a lack of appropriate instructional materials.

The term 'agroforester' was not officially recognized by many professional bodies. The concern was expressed that to offer a degree in such a 'new' subject could impede rather than help the recipients up the ladder of professional attainment. The development of agroforestry into an experimental science that can be taught, and result in the provision of adequate numbers of competent, professionally trained personnel, must be done in a way that recognizes existing professional links and existing professional standards and requirements.

Where the approach to education on the development of land use was already being taught in an integrative way, new and separate educational programs in agroforestry were not deemed necessary. Where a whole program on agroforestry education was considered desirable, or essential, then the place to start was at the postgraduate level in the form of a one or two year M.Sc. degree that combines taught causes and field work.

The greatest development of agroforestry was occurring in the tropics and subtropics; therefore, it was recommended that most postgraduate courses be situated in faculties or institutes in developing countries.

Agroforestry technologies are highly location specific, both in terms of the species used and in terms of the detailed design and management of actual operations systems. Practical course work, therefore, needs to be carried out in relation to site-specific and problem-oriented situations if it is to be fully relevant.

Short courses of various kinds were recommended on a regional basis for agroforestry awareness and buildup. Many organizations were identified who may organize these. The workshop identified the need to establish an 'International Course in Agroforestry', either at central institutions or at several regional institutions.

The materials needed for teaching agroforestry must cover both principles and practice. Written matter is available from all the important scientific subject areas to underpin the theoretical foundations for understanding and evaluating agroforestry land-use systems and practices. The task of interpretation and selection was, however, harder in some areas (socioeconomics, crop physiology) than in others (soil science). Qualified groups of specialists were deemed necessary to prepare digests of information within each discipline. Needs were identified for an inventory of existing agroforestry systems that describe the systems and provide actual data for comparison and evaluation, (e.g., in-depth case studies and visual-aid (slide) packages prepared together with explanatory booklets to illustrate the concepts, complexity and diversity of agroforestry systems around the world).

The 'twinning' of appropriate institutions in developing and developed countries was considered as an approach that may positively assist in the development of agroforestry education and training programs.

The workshop recommended that the work started for the ICRAF/DSE workshop be continued with the help of regionally appointed institutions to collect and disseminate information on agroforestry education on a regular basis. The publication of journals or gazettes on agroforestry was also encouraged.

Funding support has come slowly to agroforestry education. Donor support is essential for the implementation of the foregoing and should, therefore, be actively sought by all those involved.

The present

The current status of agroforestry in education and training programs is organized in two parts. Agroforestry in degree programs (undergraduate and postgraduate) is presented first, and agroforestry in nondegree training programs follows.

Agroforestry in degree programs

A survey of educational institutions was conducted in October 1987 to inventory existing postgraduate programs in Africa, Australia, Europe, India, and North America. More than 60 institutions were contacted and information requested on degree programs, particularly at the M.Sc. level, in forestry, agriculture, animals sciences, and other related disciplines. Also, we wanted to know to what extent agroforestry or related subject matter is being taught and whether agroforestry research is ongoing or planned. Forty-three institutions responded.

The survey complemented information already available to ICRAF on agroforestry professional education. It also showed that the systematic gathering and updating of information on where and how agroforestry is incorporated in existing education programs worldwide, is a task that requires larger human and financial resources than are available at present in a single institution like ICRAF.

Today, agroforestry or related subject matter has been incorporated in teaching and research programs at an unprecedented level. Agroforestry is found as an option for specialization in the last year(s) of B.Sc. or diploma programs in forestry, natural resource management, environmental sciences and in courses and special seminars. Full undergraduate and postgraduate programs in agroforestry are being formulated and implementation has started in quite a few. Additionally many students are choosing agroforestry-oriented research projects for their dissertations.

The scope and content of agroforestry in education programs are, however, difficult to assess from program descriptions or standard lists of course offerings. The institutional 'niches' or approaches adopted to incorporate agroforestry in existing programs is almost as varied as the number of institutions themselves. What can be recommended to assess the extent to which agroforestry is incorporated in existing programs?

A good setting for higher degree training in agroforestry aspects requires well-equipped and well-staffed combined faculties of at least agriculture, animal science, and forestry. It also requires a strong faculty commitment to a farming systems approach and the concomitant inter-departmental cooperation in teaching and research. Furthermore, agroforestry research projects that meet the requirements of postgraduate students have to be completed in nine months or up to two years. This kind of project is difficult to design except when it is a constituent part of research programs of greater amplitude and of an interdisciplinary nature. Such programs have to be conducted by a team of permanent staff of the institution who, at the same time, act as joint supervisors for the student projects. Are these elements present in existing programs where agroforestry has been incorporated?

Several points emerge for discussion from the analysis of information on agroforestry in degree programs. First, traditional forestry programs seem to be broadening the scope of the discipline. Although still concerned, as they probably should be, with the protection and/or exploitation of natural and planted tree communities, programs are beginning to reflect an understanding that a new kind of tree specialist is needed who is trained to think in terms of integrated land-use systems (rather than in terms of forests). The trend becomes clear if we take a look at institutional changes from traditional forestry departments to departments of 'Forestry and

Natural Resources', 'Forestry and Resource Management', 'Environmental Forestry', 'Horticulture and Forestry', 'Forestry and Range Management', etc.

Second, agroforestry is finding the proper 'niche' in agriculture. Agriculturists have always considered trees as perennial crops. Now they are also beginning to recognize trees as soil improvers and soil protectors, producers of fodder, food and fruit, and as producers of wood for fuel, building, fencing and other domestic and commercial purposes. This long neglect has pervaded all subjects in the curricula as well as textbooks and journals, from 'agricultural botany' to 'farming systems', 'agricultural economics' and 'rural sociology'.

Third, the role of the agroforester as a professional 'integrator' in multidisciplinary research and development projects is becoming clearer and the trend is likely to become even stronger as integrative disciplines (e.g., ecology, range management, farming systems) are incorporated into existing programs.

Fourth, six years ago full programs in agroforestry were only found in nondegree training programs, whereas today full programs in agroforestry are being developed at undergraduate and postgraduate levels. A summary of institutions offering degree programs where agroforestry is included in the academic syllabus or research programme is in Appendix 1. The faculty/department or special program under which agroforestry is taught is also indicated.

Finally, institutional structures are being created to allow for educational programs with course work and research projects spanning many disciplines. Several examples are presented in Appendix 2.

Agroforestry in nondegree training programs

Short-term training leading to the upgrading of professional knowledge and skills in agroforestry has seen in the last few years an upsurge of activities equal to if not larger than that in education. And, as in education, keeping updated records on courses, workshops, conferences, etc., worldwide, or even on a regional basis, is no small task.

In the absence of staff trained in agroforestry, the first requirement is to provide in-service training, particularly for teaching staff and research workers but also for policy-makers, planners and higher administrators. This must be done through short courses.

ICRAF attempts to inventory short-term, in-service courses organized on a regular basis (e.g., annually) by government institutions, regional or international centers where agroforestry is either the main topic or part of a larger theme. A summary of information on courses offered in 1988 that are also expected to be held in 1989 and, probably, onward is in Appendix 3A.

International Agricultural Research Centers (IARCs) and regional organizations and universities, in both developing and developed countries, play an important role in conducting training programs in support of National Agricultural Research Systems (NARS) in the developing world. They can also assist national programs to organize and conduct their own training courses in situ.

An IARCs' Workshop on Human Resource Development Through Training was held at the International Potato Center (CIP) in September 1988, in Lima, Peru, to address training issues that if addressed collectively can improve the quality and effectiveness of human resource development efforts. A digest on some issues and/or constraints related to agroforestry training follows.

First, commodity-based, on-farm research is inadequately addressed for the needs of NARS and requires a holistic approach that goes beyond the mandates of the commodity centers. The diversity of approaches and methods offered to NARS in training have had a negative impact on both the credibility of Centers and of the process. Centers can coordinate on-farm research training activities with the participation of those working in the same geographical regions. To initiate, plan, and coordinate this effort, lead centers need to be identified. Several meetings have taken place in 1988 to coordinate IARCs' networking activities, including training, operating in East and Southern Africa (e.g., ICRAF, CIAT*, ILCA*, IITA*, CIMMYT*, and others)[Summary Report 1988].

Second, disciplinary-oriented topics that cut across the research of many of the centers can be addressed collaboratively by those centers that have ongoing programs and, therefore, the comparative advantage to do so. Such courses are at present already being offered: agroforestry by ICRAF, agroclimatology by ICRISAT*, educational technology by IRRI*, and fertilizers by IFDC*. An inventory of such courses is needed for distribution to all centers and interested national institutions [Summary Report 1988].

Third, in some countries national agricultural universities are part of the NARS whereas in others they are not. Some IARCs have evolved strong linkages with these universities, including agreements for thesis research to be conducted at the centers, contracted research at the universities, and faculty participation in center training events. A common approach could

*Editors' note: These acronyms refer to various International Agricultural Research Centers.

be explored that is mutually advantageous for establishing or strengthening collaborative relationships with national agricultural universities.

Last, areas of collaboration in training materials need to be identified that can jointy be addressed by centers, universities and other national institutions. ICRAF is at present conducting an inventory of training materials in agroforestry or related subject matter in East and Southern Africa. A workshop on 'training materials in agroforestry' is tentatively planned for 1989.

ICRAF's approach to human resource development in agroforestry

Following the recommendations of the External Review that took place in 1985, a change in program focus and priorities took place recognizing the difficulty, if not impossibility, for the council to retain the overall aim of having a global impact with very limited resources. While maintaining a responsibility in continuing to develop the agroforestry discipline, the emphasis in the program of work concentrated on dissemination and technology generation through collaborative projects with national and international programs with a zonal agroecological focus in Africa – the Agroforestry Research Networks for Africa (AFRENA). An integrated 'research and training' approach was also formulated then as a more effective way of supporting agroforestry research networking and strengthening national research capabilities [Zulberti 1987].

Short in-service courses, workshops, and residential training were the initial thrusts of the training component at ICRAF. Postgraduate education started in 1988 together with the organization of specialized field trips. A few comments on focus and program development follow.

Training courses

Following the recommendations of the Nairobi workshop, ICRAF launched in 1983 a series of international courses on 'Agroforestry Research for Development: Concepts, Practices and Methods'. The model of a three-week course was developed and tested during 1983–85. The course aim is to strengthen the capability of research scientists in developing countries to initiate and implement agroforestry research leading to the generation of technologies that are both suited to local conditions and adoptable by farmers. The program focus is on the ICRAF-developed Diagnostic and Design Methodology to undertake the interdisciplinary identification of land-use problems and potentials and establish priorities for research, development and testing of sound agroforestry.

The course program is organized in modules. The scope and content of modules have been periodically revised to incorporate, change and/or adapt according to progress made in the development of the discipline. For example, the first courses offered in 1983–84 had an emphasis on agroforestry system 'descriptions', because at that time little 'quantitative data' were available. The concept of agroforestry was defined with a focus on 'systems approach to land-use management', and yet the components were mostly presented as individual units, (e.g., tree, crop, animal and man) where information was plentiful without getting too much into the interaction of these components where information was scarce. But systems theory is a tool; we did not believe that the mere addition of technical information on this subject would promote a proper understanding of land management. Furthermore, subject-oriented classificatory structures which may be helpful in the organization of scientific inquiry are not, necessarily, the best for training people on a systems approach [Huxley 1982]. Nowadays, the course program is built around the agroforestry concepts with plenty of examples and case studies, complete with data (Appendix 3B).

To date, 10 international or in-country courses have been organized. Eight of them were offered in English, one in French and English, and one in Spanish. The 'D & D course', as it came to be called at ICRAF, is offered in-country to collaborative national institutions linked to the AFRENAs during the planning phase of the networks. When agroforestry research implementation begins, then the training focus changes from the 'D & D' to 'experimental methods' and 'techniques for field research'. The course is also organized every year in May at ICRAF headquarters for a world-wide audience. Information on date, venue, sponsor, and course participants from 1983 to 1988 is in Appendix 3C. A total of 259 professionals have participated. They have come from national agriculture and forestry institutions in 51 countries in Africa, Asia, the Pacific and Latin America; about 10% of participants were women. The distribution of participants by geographical regions and countries is in Appendix 3D.

Workshops, seminars and conferences

Creating opportunities for interaction among the different parties connected with agroforestry research is an important element in developing research capability. Workshops, seminars and conferences are, broadly speaking, organized to assess problems, develop guidelines or formulate joint plans. They are usually of short duration (two to five days). In 1988, ten workshops, technical seminars and conferences took place. Six of them were held in Kenya and the rest in Rwanda (two), Malawi, and Ethiopia.

On-the-job training

The scheme's aim is to provide professionals from national institutions in developing countries with an opportunity to undertake agroforestry study/work under the supervision of ICRAF's scientific staff. Internships are six months long. The program emphasizes a practical 'learning by doing' approach, with little lecturing. Twenty-five nationals from Africa have participated in this program since 1982.

Research fellowships/visiting scientists

The program started in 1983 to allow professional staff and senior scientists from national institutions in developing countries to undertake agroforestry research alongside ICRAF scientific staff. It is a nondegree program that may last up to 24 months. Eight scientists have participated in the program from countries such as Ghana, India, Kenya, Peru, the Philippines, Tanzania and Uganda.

Postgraduate education

ICRAF collaborates with AFRENA institutions in the identification of human resource development needs and coordinates a fellowship program for postgraduate education at the M.Sc. level. The aim is to upgrade the professional qualification of multidisciplinary national research cadres so that a constant supply of well-trained researchers is guaranteed to plan and implement agroforestry research. In 1987, a five-year postgraduate education fellowship program was approved by the Canadian Agency for International Development (CIDA) for AFRENA Southern Africa (S.A.). The survey of existing postgraduate programs mentioned in this paper in the section on 'The Present', was conducted by ICRAF to identify precisely appropriate postgraduate education programs for researchers from Malawi, Tanzania and Zambia.

The fellowship program is organized in such a way that a three-person, multidisciplinary team representing food/fodder crop agronomy, forestry/horticulture, and farming systems is trained for each country. Although course work will initially be done abroad, theses research will be undertaken 'in-zone'.

The first AFRENA S.A. researchers have been admitted to start postgraduate studies in 1988 at the University of Florida and Michigan State University in the US, and McGill University in Canada. Plans are underway to start a similar program for AFRENA Eastern Africa. A project proposal was prepared in 1988 and submitted for donor support to allow 12 researchers from Burundi, Kenya, Rwanda and Uganda to undertake M.Sc. studies over a five-year period starting, hopefully, in 1989.

There are a number of institutions in Africa, as previously mentioned, which have already started or are planning to start agroforestry education programs. A proposal to start a postgraduate M.Sc. program in the region – the Agroforestry Postgraduate Program for Africa (AFPA) – was prepared by ICRAF and submitted in 1988 for donor support. It is proposed that the program be established in partnership with an African university and implemented with the participation of one or more educational institutions overseas, ICRAF and other IARCs or regional centers.

The effect of education and training programs in human resource development in agroforestry will ultimately be seen in what NARS accomplish in the generation of appropriate agroforestry technology. In the meantime, ICRAF will continue to strengthen the capability of national institutions through education and training activities in those areas of expertise where the Council has comparative advantages and available resources. The conclusions and recommendations of the present workshop will, no doubt, give a steady direction for the future of agroforestry education and training programs worldwide.

The future

It is difficult, of course, to speculate about the future of agroforestry education and training. There are, however, some issues that can probably be mentioned with reasonable certainty that will need to be tackled in the near future:

1) The number of national and international research and development institutions that take up agroforestry or agroforestry-related activities will increase rapidly over the coming few years; as a result, requests for agroforestry education and training will also increase. Are existing education and training institutions prepared to meet the demand and at the same time maintain a high-quality standard of instruction and research? Are available resources, (e.g., human, physical, financial) available and appropriate?

2) Related to the foregoing, there are several examples of professional careers that evolved in the past as a result of felt needs, but these needs were not assessed beforehand in terms of number of professionals required for that particular field, which resulted in a higher number of graduates than professional employment available. How many agroforesters are needed

today – or in the next five to 10 years? How can the supply/demand balance be best achieved?

3) There will be a rapid increase in terms of technology results coming out of research and development projects that will generate large amounts of information. Who will collect and collate information and prepare digests of use as training materials? What resources are required/available?

4) Most education and training institutions will succeed in developing institutional structures that allow for multidisciplinary programs like agroforestry; national research institutions and donor agencies alike may, however, take longer to realize that the potential of agroforestry can only be achieved through an integrated institutional approach. Are there ways/strategies to shorten this gap in time?

References

Contant RB (1980) Training and education in agroforestry. *In*: Chandler T and Spurgeon D (eds.). International cooperation in agroforestry: Proceedings of an international conference, pp. 191–218. ICRAF, Nairobi

Huxley PA (1980) Agroforestry at degree level: a new program structure. *In* Chandler T and Spurgeon D (eds.). International cooperation in agroforestry, pp. 219–227; ICRAF, Nairobi

Huxley PA (1982) Education for agroforestry. A discussion paper presented at the United National University Workshop on Agroforestry, Freiburg, Germany. 31 May–5 June

Lundgren B (1987) Institutional aspects of agroforestry research and development. *In*: Steppler HA and Nair PRK (eds). Agroforestry: A decade of development, pp 43–66. ICRAF, Nairobi

Schuh EG (1987) Policy and research priorities for agroforestry. Paper delivered at the 10th Anniversary Conference of ICRAF, 7–11 September, 1987, Nairobi

Steppler HA and Nair PKR (eds)(1987) Agroforestry: A decade of development. ICRAF, Nairobi

Summary Report (1987) International Agricultural Research Centers' workshop on human resource development through training, 12–16 September, 1988. International Potato Center, Lima

Zulberti E (ed)(1987) Professional education in agroforestry: Proceedings of an international workshop. ICRAF, Nairobi

Zulberti E (1987) Agroforestry training and education at ICRAF: Accomplishments and challenges. Agroforestry Systems 5: 353–374

Appendix 1

Agroforestry in degree programs (full address of insitutions is available at ICRAF on request)

(i) at undergraduate B.Sc. or Diploma level
Forestry – Australian National University (Australia), University of Toronto (Canada), Moi University (Kenya), Egerton University (Kenya), Gadjah Mada University (Indonesia), University of the Philippines in Los Banos (Philippines), University of British Columbia (Canada);
Forestry and Resource Management – University of California in Berkeley (USA), University of Ibadan (Nigeria);
Forestry and Natural Resources – Edinburgh University (UK);
Social Forestry – Institute Pertanian Bogor (Indonesia), Universiti Pertanian Malaysia (Malaysia);
Forest and Wood Sciences – Colorado State University (USA);
Agriculture – University of Guelph (Canada);
Agriculture and Forestry – Escuela de Ciencias Ambientales (Costa Rica), Universidad Nacional Heredia (Costa Rica), Instituto Tecnologico Cartago (Costa Rica), Facultad de Agronomia (El Salvador), Universidad de San Carlos (Guatemala), Universidad Saldivar (Guatemala), Facultad de Agronomia y de Ciencias Forestales (Honduras), Universidad Autonoma (Mexico), Colegio Superior de Agricultura Tropical (Mexico), University of Aberdeen (UK);
Management Development Programme for Africa – Silsoe College (UK);
Veterinary Medicine and Animal Science – Universiti Pertanian Malaysia (Malaysia);
Biology – Wau Ecology Institute (Papua New Guinea), Universidad Centro Americana (Nicaragua);
Botany and Horticulture – University of Hawaii (USA);
Agroforestry – Memorial State University (Philippines), University College of North Wales in Bangor (UK);
Natural Sciences – Facultad de Ciencias Ambientales (Guatemala).

(ii) in postgraduate M.Sc. or Diploma programs
Forestry – Australian National University (Australia), University of Melbourne (Australia), Facultade de Ciencias Agrarias do Para (Brazil), Universidade Federal Rural de Pernambuco (Brazil), Universidad de Santiago (Chile), Universidad de Caldas (Colombia), University of New Brunswick (Canada), Moi University (Kenya), University of Cantebury (New Zealand), University of Technology (Papua New Guinea), Kasetsart University (Thailand), Universidad Agraria La Molin (Peru), Michigan State University (USA), University of Kentucky (USA), University of Massachusetts (USA), Oregon State University (USA), University of Illinois (USA), Texas A&M University (USA);
Forest Management – Ibadan University (Nigeria);
Forestry and Environmental Studies – Yale University (USA);
Forestry and Natural Resources – Colorado State University (USA);
Forestry and Resource Management – University of California, Berkeley (USA);
Forest Resources and Conservation – University of Florida (USA);
Tropical Forestry – Wageningen Agricultural University (the Netherlands);
Forestry and Range Management – Washington State University (USA);
Forestry for Rural Development – International Institute for Aerospace Survey and Earth Sciences (ITC) and the International Agricultural Centre (IAC)(the Netherlands);
Community Forestry in Rural Development – same as above;
Social Forestry – University Pertanian Malaysia (Malaysia), University of Reading (UK), Oxford University (UK);
Environmental Forestry – University College of North Wales, Bangor (UK);
Environmental Studies – University of the Philippines at Los Banos (Philippines), University of East Anglia (UK);

Horticulture and Forestry – University of Horticulture and Forestry, Solan (India);
Agriculture and Forestry – University of Aberdeen (UK);
Agricultural Development – Technical University of Berlin (Germany);
Agriculture – University of Maine (USA), University of Guelph (Canada);
Agricultural Sciences and Natural Resources – Universidad de Costa Rica/Centro Agronomico Tropical de Investigacion y Ensenanza (CATIE)(Costa Rica);
Renewable Natural Resources – University of Science and Technology, Kumasi (Ghana);
Management Development Programme for Africa – Silsoe College (UK).

Appendix 2

Some examples of institutional structures created to allow for educational programs spanning several disciplines

(1) The University of Alberta has a unique Faculty of Agriculture and Forestry in Canada where two land-based disciplines are both taught as Faculty programmes and share some of the common basic subjects. Although the Faculty offers no formal course in agroforestry as such, it is able to provide synthesis courses in this subject through directed studies and research work.

(2) The University College of North Wales (UK) formed the School of Agricultural and Forest Sciences with a land-management bias. In addition, the Centre for Arid Zone Studies was created within the framework of the School to deal with problems of land use in difficult environments in the semi-arid and arid tropics. A 3-year B.Sc. degree program in agroforestry is offered. At the postgraduate level, there is no specialized formal agroforestry teaching, but there are research opportunities.

(3) The University of Melbourne is at present the only one in Australia in which both agriculture and forestry are represented. Research in agroforestry has been proceeding for some time. Having developed a strong research base, the Faculty has now moved to develop graduate training in agroforestry. A course work programme cutting across disciplines has been developed with particular strengths in agroforestry. Students who achieve a satisfactory result in the postgraduate diploma course work can transfer to complete a Master's degree by research (in 2 years' time).

(4) Also in Australia, the Department of Forestry at the Australian National University will merge with the Department of Geography in 1989 to become the School of Resources and Environmental Management where postgraduate (and undergraduate) courses in agroforestry will continue to be taught. Agroforestry research is actively pursued by the Forestry Department and the Centre for Forestry in Rural Development in Australia, Africa, Asia, and the Pacific.

(5) The Center for Semi-Arid Forest Resources at the Caesar Kleberg Wildlife Research Institute of Texas A & I University in the US conducts research leading to the improvement of valuable plants for arid and semi-arid regions. Information obtained from these studies is being applied in the development of international agroforestry programs. Even though no formal agroforestry courses are available, M.Sc. degrees are offered by the Department of Agriculture with thesis research in agroforestry.

(6) The International School of Forestry and Natural Resources at Colorado State University in the US offers nondegree and degree programmes at M.Sc. and Ph.D. levels in forestry where course work includes forestry, horticulture and agronomy as well as specialized courses in agroforestry.

(7) The Yale Tropical Resources Institute (TRI), created in 1984, provides a focus for tropical resource studies and training at Yale University in the US. TRI has a broad-based curriculum in tropical natural resources research and management. Courses in tropical forestry typically include tropical economic botany, tropical ecology, rural development sociology, and tropical natural history. Students have access to classes in the departments of Anthropology, Economics, Political Science, Sociology, the School of Organization and Management, and others. ICRAF is a member of TRI's Advisory Group.

(8) At the University of Florida, the School of Forest Resources and Conservation is a unit of the College of Agriculture. The primary objective of the School's three Departments – Forestry, Wildlife and Range Sciences, and Fisheries and Aquaculture – is to provide professional education in the areas of forestry, wildlife ecology, and resource conservation. An Interdisciplinary Program in Agroforestry was started in 1987, which offers graduate education at Masters and Ph.D. levels and short-term training courses in agroforestry. Additionally, UF offers courses in Farming Systems Research/Extension, Food in Africa, Tropical Forestry, and other issues related to agricultural development and natural resource management where agroforestry elements are included.

(9) At the University of Ibadan in Nigeria, the Faculty of Agriculture and Forestry offers agroforestry at the undergraduate level in the Department of Forest Resources Management as a three-unit course during the 4th year, which is also a practical year. In addition, introduction to land-use planning is taught in the 3rd year and multiple land use is offered in the 5th year. The Department of Forest Resources Management has been involved in agroforestry research for some time.

(10) The Institute of Renewable Natural Resources at the University of Science and Technology in Kumasi was established in 1982 to promote the proper management and utilization of forests, savannas, wildlife, freshwater fisheries and watersheds through teaching, research, and extension. Four departments representing the foregoing areas have been involved in offering courses leading to B.Sc., Diploma and M.Sc. degrees in 'renewable natural resources' and 'wood technology and industrial management'. Courses are designed to train graduates in a multiple-use management approach based on the ecosystem concept 'to plan for the whole rather than any one particular resource and be able to discuss intelligently with and seek advice from other specialists' (personal communication with Asare, 1984). Agrosilvopastoral research integrating tree crop, food crop and small ruminants has been conducted. A postgraduate diploma course is agroforestry started in 1988.

Appendix 3A

Agroforestry in nondegree training programs: Training courses in AF or related subjects offered regularly by national and international organizations

Venue	Organizer	Duration	Expected audience	Title/topics
Tel Aviv Israel	Centre for International Agricultural Development Cooperation (CINADCO) Ministry of Agriculture P.O.B. 7054 Tel-Aviv 61070 Israel Tel: 03-211490/492 Tlx: 361496 MINAG IL	3 months	Government and non-government officials and extension workers who have worked in agriculture, pasture and agroforestry in deserts, arid and semi-arid zones for at least 3 years	Agroforestry, Desert Agriculture and Extension *Program*: intensive irrigated agriculture under desert conditions, soil and water management, agroforestry under rainfed conditions, etc.
Nairobi Kenya	International Council for Research in Agroforestry (ICRAF) P.O. Box 30677 Nairobi Kenya Tel: 521450 Tlx: 22048	3 weeks (every May)	Research scientists in agriculture, forestry, livestock science, social sciences	Agroforestry Research for Development *Program*: concepts and practices of agroforestry: farming systems and agroforestry; diagnosis of LUS, evaluation of agroforestry systems, experimental designs
University of the Philippines at Los Banos, College, Laguna PHILIPPINES	Winrock International F/FRED Training Unit P.O. Box 1038 Kasetsart Post Office Bankhen, Bangkok 10903	2.5 weeks	Foresters	Social Science Concepts and Methodologies for Foresters *Program* social sciences research methodologies for Foresters
		2.5 weeks	Foresters and social forestry officers	*Program* forestry, forestry systems and people oriented forestry

Appendix 3A *Cont*

Venue	Organizer	Duration	Expected-audience	Title/topics
Wageningen The Netherlands	International Agricultural Centre (IAC) P.O. Box 88 6700AB Wageningen The Netherlands Lawickse Allee 11 Telegrams: INTAS Tel: 08370-90111 Tlx: 45888-INTAS NL	3 months	Policy staff in developing countries from governments and non-government organizations. Preference to teams of two participants per country, one forester and one agronomist/ livestock or rural planning	The Design of Community Forestry *Program* forestry in rural development, ecology, sylviculture, forest survey technology, agroforestry, watershed management, energy, land evaluation, etc.
Bedford United Kingdom	Silsoe College Professional Development Center Silsoe College Silsoe Bedford MK45 4DT Tel: Silsoe (0525) 60428 Tlx: 265871 (MONREFG) REF EVM305	1 month	Technical staff working in soil conservation in third world countries	Soil Conservation *Program* soil conservation design, soil conservation for sustainable farming or land husbandry systems
Edinburgh United Kingdom	TROPAG Consultants Ltd. The School of Agriculture University of Edinburgh West Maires Road Edinburgh Scotland, UK EH9 3JG Tel: 031-667 1041 Fax: 031-667 260 Tlx. 727617	2 months	Research scientists and extension staff	Tropical Agroforestry *Program*: agroforestry, agriculture, forestry and extension methods for the introduction or improvement of agroforestry systems

West Midlands United Kingdom	Agricultural Education and Training Unit (AETU) The Polytechnic Wolverhampton Castle View Dudley DY1 3HR West Midlands Tel: 0384-459741 Tlx: 336301 POLWOL G	3 months	Teachers and trainers of agroforestry, forestry and other natural resources subjects	Agroforestry/Forestry Teaching *Program*: teaching methodologies for natural resource subjects
Fort Collins Colorado USA	International School of Forestry & Natural Resources College of Forestry & Natural Resources Fort Collins, Colo 80523 Tel: (303) 491-5443 Tlx: 9109309011	Various	Government officials or private employees working or teaching on forestry and natural resources	Various topics, e.g. – natural resources – natural resources ecology – range science – forest and wood sciences – earth resources, etc.
Logan Utah USA	Utah State University Range Science Department Logan, Utah USA 84322-5005 Tel: (801) 750-1696 Tlx: 3789426	3 weeks	Students, administrators and land managers	Desertification, Rehabilitation and Management of Rangelands in Pastoral Systems *Program*: pastoral production systems, ecology of arid lands, range/livestock extension, etc.

Appendix 3A *Cont*

Venue	Organizer	Duration	Expected-audience	Title/topics
Logan Utah USA	International Irrigation Center and Utah State University Department of Agricultural and Irrigation Engineering Utah State University Logan, Utah 84322-4150 USA	5 weeks	Agricultural Engineers Watershed Management Specialists, Conservationists, and Agronomists	Soil & Water Conservation Management *Program*: erosion processes and measurement, ecnomics and soil and water conservation, biological treatments (agroforestry), etc.
Gainesville Flordia USA	Department of Forestry and Training Division IFAS University of Florida Gainesville, FL 32611 USA	5 weeks	Agricultural and Forestry Professionals from developing countries	Agroforestry Extension & Training *Program*: includes practical field training in a developing country

Appendix 3B

Modules for the ICRAF course to be held 8–26 May 1989

I – Introduction to agroforestry

The systems approach, agroforestry systems and practices, multipurpose trees and shrubs for agroforestry systems, agroforestry systems and their recommendation domains, and interaction of components in agroforestry systems;

II – Diagnosis and design

Introduction to analysis of ecozonal land-use systems (macro and micro) to identify constraints and agroforestry opportunities, identification and specification of agroforestry improvements or interventions, ex ante analysis or proposed agroforestry systems/technologies for existing farming systems, and derivation of research and extension agenda;

III – Agroforestry experimentation

Technology development process, and critical steps in design and analysis of on-station research and on-farm experimentation, e.g., identification of researchable problems, definition of experimental objectives/hypotheses, formulation of treatments/treatment combinations, selection of experimental designs;

IV – Agroforestry evaluation

Main evaluation domains: ecology, biology, economy, sociology; time frame for evaluation; user-perspective evaluation; evaluation of agroforestry components, practices or systems; and impact on other systems.

Appendix 3C

ICRAF training courses 1983–1988 and participants

	I	II	III	IV	V	VI	VII	VIII	IX	X	Total
Venue	ICRAF Kenya	ICRAF Kenya	UPM Malaysia	UNIPA Peru	ICRAF Kenya	Hyderabad India	Chipata Zambia	ICRAF Kenya	ICRAF Kenya	ICRAF Kenya	–
Dates	1–18 November 1983	4–22 June 1984	1–19 October 1984	3–22 June 1985	4–22 November 1985	16–30 September 1986	Dec 5–17 November 1986	11–29 May 1987	Nov 23–Dec 10 1987	9–27 May 1988	–
Sponsors	ICRAF/USAID	ICRAF/USAID	ICRAF/USAID	ICRAF/USAID	ICRAF/USAID	ICRAF/Ford Foundation	ICRAF/IDRC	ICRAF/SIR	ICRAF/USAID	ICRAF/DSO	–
Participants											
A. Africa	22	21	0	0	28	0	24	12	22	20	149
B. Asia & Pacific	0	0	19	0	1	29	0	10	0	15	74
C. Latin America	0	3	0	26	0	0	0	3	0	4	36
Total	22	24	19	26	29	29	24	25	22	39	259

Appendix 3D

Participants to ICRAF training courses

Region	Country	Courses										
		1 1983	II 1984	III 1984	IV 1985	V 1985	VI 1986	VII 1986	VIII 1987	IX 1987	X 1988	Total
A. Africa	Benin	–	1	–	–	–	–	–	–	–	1	2
	Botswana	–	1	–	–	–	–	–	–	–	–	1
	Burundi	1	–	–	–	2	–	–	–	6	–	9
	Cameroun	–	–	–	–	–	–	–	–	–	1	1
	Cape Verde	1	–	–	–	–	–	–	–	–	–	1
	Ethiopia	2	–	–	–	2	–	–	1	1	4	10
	Ghana	1	1	–	–	–	–	1	1	–	1	5
	Kenya	6	5	–	–	7	–	1	1	8	1	29
	Lesotho	–	–	–	–	–	–	–	–	–	1	1
	Liberia	–	1	–	–	–	–	–	–	–	1	2
	Malagasy	1	–	–	–	3	–	–	–	–	–	4
	Malawi	1	1	–	–	1	–	4	–	–	–	7
	Mali	–	–	–	–	1	–	–	–	–	1	2
	Mauritius	1	–	–	–	–	–	–	–	–	–	1
	Morocco	–	–	–	–	–	–	–	–	–	1	1
	Mozambique	–	–	–	–	–	–	–	–	–	1	1
	Niger	1	–	–	–	–	–	–	–	–	–	1
	Nigeria	1	2	–	–	1	–	–	1	–	–	5
	Rwanda	–	–	–	–	2	–	–	–	1	1	4
	Senegal	–	1	–	–	–	–	1	–	–	–	2
	Sierra Leone	–	–	–	–	–	–	–	2	–	–	2
	Somalia	–	–	–	–	–	–	–	1	–	1	2
	Sudan	–	2	–	–	5	–	–	3	–	–	10

Appendix 3D (Continued)

Region	Country	Courses										Total
		1 1983	II 1984	III 1984	IV 1985	V 1985	VI 1986	VII 1986	VIII 1987	IX 1987	X 1988	
	Tanzania	3	1	–	–	–	–	4	–	–	1	9
	Tunisia	–	–	–	–	–	–	–	–	–	1	1
	Uganda	2	3	–	–	3	–	–	1	6	1	16
	Zaire	–	–	–	–	–	–	–	1	–	1	2
	Zambia	1	–	–	–	–	–	9	–	–	–	10
	Zimbabwe	–	2	–	–	1	–	4	–	–	1	8
	Sub-total	22	21	–	–	28	–	24	12	22	20	149
B. Asia & Pacific	Bangladesh	–	–	–	–	–	–	–	–	–	4	4
	China	–	–	–	–	–	–	–	–	–	2	2
	India	–	–	1	–	–	29	–	4	–	1	35
	Indonesia	–	–	4	–	–	–	–	–	–	–	4
	Malaysia	–	–	6	–	–	–	–	1	–	1	8
	Nepal	–	–	–	–	–	–	–	1	–	1	2
	Pakistan	–	–	–	–	–	–	–	–	–	1	1
	Papua New Guinea	–	–	–	–	–	–	–	–	–	1	1
	Philippines	–	–	4	–	1	–	–	1	–	1	7
	Sri Lanka	–	–	–	–	–	–	–	–	–	1	1
	Thailand	–	–	4	–	–	–	–	1	–	2	7
	Vietnam	–	1	–	–	–	–	–	2	–	–	3
	Sub-total	–	–	19	–	–	29	–	10	–	15	74

C. Latin America and Caribbean	Belize C.A.	–	–	–	–	–	–	–	–	–	1	1
	Bolivia	–	–	–	3	–	–	–	–	–	–	3
	Brazil	–	–	–	4	–	–	–	–	–	1	5
	Colombia	–	–	–	4	–	–	–	–	–	–	4
	Costa Rica	–	1	–	–	–	–	–	1	–	–	2
	Ecuador	–	–	–	3	–	–	–	–	–	1	4
	Guatemala	–	–	–	–	–	–	–	1	–	–	1
	Peru	–	2	–	9	–	–	–	–	–	1	12
	Venezuela	–	–	–	3	–	–	–	–	–	–	3
	Trinidad	–	–	–	–	–	–	–	1	–	–	1
	Sub-total	–	3	–	26	–	–	–	3	–	4	36
	Total No. of participants	22	24	19	26	29	29	24	25	22	39	259

Appendix 4

Survey of educational institutions contacted in Africa, Australia, Europe, Indian and North America

A. AFRICA
Dean
Forest Resource and Wildlife Management
Moi University
P.O. Box 3900
Eldoret
KENYA

Egerton University
Private Bag
P.O. Njoro
Nakuru
KENYA

Head
Department of Forest Resources Management
University of Ibadan
Ibadan
NIGERIA

Sakoine University of Agriculture
Faculty of Forestry
Department of Wood Utilisation
P.O. Box 3009
Morogoro
TANZANIA

B. AUSTRALIA
Head
Department of Forestry
Australian National University
Canberra ACT
AUSTRALIA

Prof. I. Ferguson
Faculty of Agriculture & Forestry
University of Melbourne
Parkville
Victoria 3052
AUSTRALIA

C. EUROPE
Dean
Forestry Institute 'HINKELOORD'
Wageningen Agricultural University
P.O. Box 342
6700 AH Wageningen
THE NETHERLANDS

Principal
University College of North Wales
Bangor
U.K.

Vice Chancellor
University of Oxford
University Offices
Wellington Square
Oxford, OX1 2JD
U.K.

Vice Chancellor
University of Edinburgh
Edinburgh, EH8 9YL
U.K.

Vice Chancellor
University of Aberdeen
Aberdeen, AB9 1FX
U.K.

Dean
Faculty of Agriculture
University of Reading
White Knights
P.O. Box 217
Reading RG6 2AN
U.K.

Silsoe College
Silsoe
Bedford, MK45 4DT
U.K.

D. INDIA
Registrar
University of Horticulture and Forestry
Solan1-173 230
INDIA

E. NORTH AMERICA

Canada
Dean
Faculty of Forestry
University of New Brunswick
Fredericton
New Brunswick
CANADA

Dean
Faculty of Forestry
University of Toronto
Toronto
Ontario M5S 1A1
CANADA

University of Guelph
Faculty of Agriculture
Guelph
ONTARIO N1G 2W1
CANADA

University of British Colombia
Vancouver, BC V6T 1W5
CANADA

F. UNITED STATES OF AMERICA

Vice-Chancellor
University of Hawaii
3190 Maile Way
Honolulu 96822
Hawaii
U.S.A.

Head of Department
Forestry and Resource Management
University of California
Berkeley
California 94720
U.S.A.

Head
Department of Forest and Wood Sciences
Colorado State University
Fort Collins
Colorado 80523
U.S.A.

Dean
School of Forestry and Environment Studies
Yale University
New Haven
Connecticut 06511
U.S.A.

Head
Department of Forestry
University of Illinois
Urbana
Illinois 61801
U.S.A.

Department of Forestry
Michigan State University
East Lansing
Michigan 48824
U.S.A.

President
International Programs
College of Environmental Science and Forestry
State University of New York
Syracuse
New York 13210
U.S.A.

Head
Department of Forest Science
Oregon State University
Corvallis
Oregon 97331
U.S.A.

Head
Caesar Kleberg Wildlife Research Institute
Texas A & I University
Kingsville
Texas 78363
U.S.A.

Dean
Department of Forestry and Range Management
Washington State University
Pullman
Washington 99164
U.S.A.

Dean
College of Natural Resources
University of Wisconsin
Stevens Point
Wisconsin 54481
U.S.A.

Head
Department of Forestry
University of Florida
Gainesville
Florida 32611
U.S.A.

G. SOUTH EAST ASIA

Institut Pertanian Bogor
Jl. Raya Pajajaran
P.O. Box 28
Bogor Timur
INDONESIA

University of the Philippines
Los Banos
PHILIPPINES

Don Mariano Marcos State University
Bacnotan
La Union
PHILIPPINES

Universiti Pertanian Malaysia (UPM)
Serdang
Selangor
MALAYSIA

G. SOUTH AMERICA

Facultade de Ciencias Agrarias do Para
CP 917, 66000
Belem
BRAZIL

Universidade Federal Rural de Pernambuco
Rua D. Manoel de Medeiros s/n
Dois Irmaos, 50000 Recife, PE
BRAZIL

Universidad de Santiago de Chile
Avda Ecuador 3469
Casilla 4637
Correo 2
Santiago
CHILE

Universidad de Caldas
Apdo Aereo 275
Manizales
Caldas
COLOMBIA

Agroforestry Systems **12**: 41–48, 1990.

Keynote paper

Agroforestry education

HOWARD A. STEPPLER
McGill University, Montreal, Canada

Abstract. Development of agroforestry education is following the pattern of evolution of some other areas of study such as plant pathology, genetics, and statistics. At universities these three areas began within another department or departments, and after being moved into their own departments began to flourish and develop their own identity. However, the main question is what can or should be done to further the process? The increasing number of agroforestry projects in the world, the lack of trained agroforesters, and the estimated increased need for agricultural scientists are all indicators for the future demand for trained agroforesters. Career opportunities for professional agroforesters lie in three areas: as research scientists, as extension agents, or as development agroforesters. Two broad educational approaches to setting educational objectives are to (1) set objectives on the basis of the perceived problems likely to be encountered in agroforestry and (2) set objectives according to roles which agroforesters are likely to assume. The design of both undergraduate and postgraduate curricula are discussed. The unique core is the agroforestry system per se and the development of a systems analysis methodology. The institutional structure and clear goal definition can facilitate the development of an agroforestry program, but in the final analysis it is the dedication and enthusiasm of the individual staff that will count for success.

Introduction

The International Council for Research in Agroforestry (ICRAF) celebrated its 10th anniversary in September 1987. The celebration activities included seminars, workshops and the publication of a book [Steppler and Nair 1987]. In the book there are several references to the need for expanded education. Two examples are: 'Educational and training programs in agroforestry are key requirements as we move into this second decade of agroforestry development' (p. 18), and 'Conservation and management of soil and water through the integration of trees must become part of the ordinary training of agricultural and other extension agents' (p. 197).

Further, the Bene Report (Bene et al. 1977), which has as its major recommendation that ICRAF be established, listed as one of the activities which the council might undertake: 'the promotion of the teaching of the principles of agroforestry at all levels in the education system'.

ICRAF also held an international workshop on agroforestry education with the proceedings published in 1987 [Zulberti 1987]. Among the pertinent papers are one by von Maydell in which he examines the question, 'What will be expected of professional agroforesters?' (p. 156–163) and the report of working group five in which is presented a listing of components of 'Knowledge Expected of Agroforesters' (p. 340). ICRAF regularly conducts training courses, but these activities have by no means exhausted the subject.

Agroforestry: a new area of study

The brochure announcing this conference cites agroforestry as a new discipline. This immediately caused me to question, 'Is it a discipline?' Additional thoughts, discussions with colleagues, and excursions into dictionaries, convinced me that to raise the issue would be non-productive as far as this conference is concerned. However, the need for a disciplined mind in agroforestry became apparent in order to avoid meandering down the various pathways of activity. The significant question is the nature of the discipline which needs to be instilled into every agroforester.

Agroforestry is a new area of study and activity, in the context of the word but not the concept. It may be instructive to consider the manner of evolution of some other 'new' areas of study. Let us briefly examine three: plant pathology, genetics, and statistics.

Plant diseases have been known about since time immemorial. Theophrastus described many in 300 B.C. Formal teaching began in Europe in the mid-19th Century with lectures associated with botany. The first chair in plant pathology was in Copenhagen in 1883.

In America, the first lectures were in cryptogamic botany in the 1870's, with the first department of plant pathology established at Cornell in 1907. Plant pathology has arisen as a study area within botany and gradually evolved to full departmental status.

Genetics followed a somewhat similar path. There are biblical references to inheritance, as in Genesis 1:11, 'grass and herb yielding seed after his kind'. Mendel's paper of 1866, rediscovered at the beginning of this century, set the basis for the development of genetics. It was, however, taught in various departments, including botany, zoology, plant breeding or animal breeding. The first department of genetics was established in Canada in 1945 at McGill University. A common practice was to have a program in genetics managed by a committee drawing members with training and programs in genetics from various departments of the university.

Probability, one of the major fields of theory upon which statistics is based, has its roots firmly in mathematics. It was not until Gosset, writing under the pseudonym of 'Student' in 1908, and R.A. Fisher's *Statistical Methods for Research Workers* in 1925, based on studies of the long-term data from Rothamsted which used the principles of probability but with samples and the concept of experimental design, that our modern statistics was developed. There is still much research and teaching in basic theories in mathematics as well as the statistical approaches to be developed in statistics departments for the use of the non-mathematical researcher.

This brief retrospective on three areas of study leads to two observations. First, each area began within another area (the reason for its initiation is not germane to this paper). Second, once the subject area had moved out from this umbrella and became institutionalized, it began to flourish and develop its own identity.

Agroforestry is following the same pattern. Foresters and agronomists developed the concept and have nurtured the idea. Institutionalization began with the establishment of ICRAF. It has begun at the university level, with developments such as at University of Florida. Other institutions have introduced courses and programs, particularly at the graduate level. The major question is what can or should be done to further the process?

This question will be examined by looking first at the demand for agroforesters, the kind of agroforesters deemed necessary, and in what broad categories they would find employment. Next, the broad problem of setting educational objectives, including the consideration of a common core and career placement of graduates will be looked at. Finally, a brief look at the institutional structure necessary to carry out the tasks will conclude, the paper.

Demand for agroforesters

There has not been a survey conducted to determine the demand for agroforesters. Consequently, one must proceed with some caution and use circumstantial evidence. The demand, at least in the developing world, will be generated by two forces: the activity of donors of assistance and their perception of need, and the problems as seen by national governments and agencies which are amenable to agroforestry solutions.

Two pieces of evidence from the donor community strongly suggest a rapidly growing demand. The number of donors who support ICRAF has grown from four in 1978–79 to 15 in 1987. All of these increasingly have agroforestry projects which they are supporting or intend to support. This creates a demand for agroforesters at the headquarters of these agencies as well as agroforesters in the field to manage the projects. Spears [1987] states that funding for agroforestry-type projects undertaken by the World Bank during 1977–1986 has grown from 6% of forestry projects to 37%.

The best evidence of demand from the client has been the experience of

ICRAF in developing its regional network programs in Africa and in responding to requests from elsewhere. The demands on ICRAF have been overwhelming. There is never the problem of insufficient projects, but always the necessity to set aside some country requests for assistance. There is one common feature to virtually all such requests: the countries do not have agroforesters or agroforestry training capabilities.

There is one further piece of evidence. The World Bank studied the agricultural research needs of East and West Sub-Saharan Africa from 1975 through 1986. Using a scenario assuming that only 0.33 percent of agricultural domestic product was invested in agricultural research, they estimated that upwards of 10 000 additional agricultural research scientists would be needed by the year 2000. There was no provision made for agroforestry.

However, if one makes a very modest assumption that 5% of these would be agroforesters and that a further 15% would be agricultural scientists working on agricultural components of an agroforestry system, then we face training 2000 professionals by the year 2000 – formidable as a task, but not unrealistic as a need. And this only addresses Africa, similar needs are developing in other parts of the world.

It should be noted that these comments, and in particular the estimates, refer only to researchers. There is no provision for training extension agroforesters, where the need will be as great if not greater.

Opportunities for agroforestry professionals

One can consider three broad areas of opportunity for agroforesters: as research scientists (technology generation), as extension agents (technology transfer), or as development agroforesters (project implementation). As a research scientist, the agroforester is liable to be working at any level of the research spectrum, from very basic studies to highly applied field-oriented research. He/she may work in any one of the many areas which fall within the gamut of agroforestry, from nutrient cycling to genetic improvement of trees, to socio-cultural impacts of new technology. Irrespective of the specific problem, we must assume that the work is in response to a rigorously defined question which has been identified as pertinent for a particular situation. In this case, the ultimate goal of the research is to develop new and appropriate technology.

The extension agent will have a dual role to play. Agents will be the key in the transfer of technology to the ultimate user and, should be involved with the testing of the appropriateness of that technology at the farmer level.

Thus, agents must be able to determine the nature of the intervention which is most timely and acceptable to the user and, be able to evaluate that relevance to the problem. Extension's role, which is intimately associated with the first, is to feed back information to the research scientist. Therefore, the extension agent is a vital link in the technology generation and transfer system and, hence, must have some common ground upon which to meet the researcher.

Those employed in the development field will require much the same skills as the extensionist, with the probable additional responsibility of management and/or design of a project. Thus, they will need the ability to insure the integration of the selected technology into the local situation. Further, they will need to monitor the ongoing process in order to be able to spot problems and seek solutions. Thus, they too must be able to communicate with the technology generators, the local transfer agents, and the donor agencies which support the projects. In preparing project proposals, they will require the same skills but in a different sequence.

Educational objectives

Educational objectives should aim to prepare practitioners of agroforestry to perform in accordance with the definition of the field [Lundgren 1982]. There are at least two broad approaches: one is to set objectives on the basis of the perceived problems likely to be encountered in agroforestry, and the other is to set them according to the roles which agroforesters are likely to assume.

The first method presents many choices, as evidenced from the definition of agroforestry [Lundgren 1982]. Multipurpose trees are the unique feature of any agroforestry system. Identification of the most appropriate out of the more than 2000 candidate species is clearly a priority problem. However, there are many other problems. For example what are the necessary environmental inputs? What spatial arrangement of a system is best? How does one identify and counter gender and/or age bias in research and extension? What is the impact of land tenure or forest policies on agroforestry development? It should be apparent that this procedure will give a strong bias to the background of the individual setting the objectives. It could readily lead to different objectives in different educational institutions for what is purported to be the same professional preparation.

The three broad roles envisaged for agroforesters have been discussed. The question, 'Is there a common denominator for these roles which could then serve as an educational objective?', must be answered. The descriptions for roles of the extension worker and for those in development indicate a common ground and suggest many of the same necessary skills. Skills include the full understanding of the agroforestry concept, with the ability to determine the nature of intervention to improve the efficiency of existing systems, the ability to facilitate integration of technology into systems, and the ability to monitor systems and suggest further modification/intervention as necessary. All this would assist the extension agent in better serving clients and in improving this feedback. A 'systems analysis' capability would, in the same manner, serve development agroforestry. I submit that if systems analysis is conducted in the manner implied for the extension agent, then one should be able to develop that rigour in defining appropriate research questions. Thus, the development of a systems analysis capability is equally essential for a researcher as for an extension agent, and becomes the unique core of the professional agroforester.

Levels of educational preparation

Programs can operate at both the undergraduate and postgraduate level. At the undergraduate level there is the possibility of designing a curriculum or major in agroforestry. The central feature of such a major would be a course or courses dealing with systems of agroforestry, methods of analysis of systems and/or field situations in which agroforestry is thought to be a potential land-use system. Case studies would be a valuable means of presenting some of this material. Since multipurpose trees are a unique feature of agroforestry, it would be most desirable to have some presentations in this area. Environmental impact, and particularly sustainability, is another important aspect of agroforestry and would warrant serious considerations for inclusion in the 'core' package. Socio-economic issues might be covered in either the systems section or the impact section or both.

The other components of the curriculum/major would depend on the expertise available. Students could develop depth in crops for agroforestry systems, animal agriculture in agroforestry systems, or product output, use and marketing from an agroforestry system, etc. The more basic biological, social and physical sciences pertinent to the subject area would also be included.

The postgraduate level is a different situation. Students entering a program can come from either an undergraduate program in which the student majored in agroforestry, or a non-agroforestry background. Students entering from the first present no problem. Their graduate program will presumably lead them to undertake research and further reading and courses in a chosen area of specialization. This will be done with the core program in agroforestry as a broad background within which the student should readily see the relevance of more specialized study.

Students entering from a non-agroforestry background present other difficulties. If the objective is to become an agroforester with an advanced degree, but the student does not have an agroforestry 'core', it would seem obvious that he/she must develop some capability in systems analysis and particularly in constraint/problem identification. The optimum situation would be for such a student to take the 'core' program identified for the undergraduate, although a truncated version may be a more realistic goal.

One skill that the postgraduate student should develop is that of problem identification and research formulation. There should be a clear understanding of the relationship between areas of specialization and the more general systems context. The student should also be able to appreciate the potential impacts of the speciality research, and the flexibility necessary to insure the greatest benefit for the ultimate user.

It is difficult to establish the level of agroforestry course program which a postgraduate student should be required to take. To put them at the same level as an undergraduate who majored in agroforestry, would strongly suggest that such students should take the same 'core' program. Should we prepare post-graduate programs in agroforestry which are non-thesis, non-research, but which would provide the 'core' subjects perhaps with more case studies), etc.? This certainly could have many attractions for someone preparing for research.

Institutional setting

One can rarely start with a clean slate and design the optimal institution. Rather, we are faced with ongoing programs and must decide how to establish agroforestry with the best chance of survival and growth and least amount of friction.

Institutionalization was and is an essential part of the development of new areas. The area must have its own identity and home base. The struggle may be mostly between forestry and agriculture. The ideal may be for agriculture and forestry to be combined into one faculty of agroforestry.

The more usual situation is to have separate faculties. Agroforestry is bound to be developed by one of these, quite probably forestry. The danger is that it takes on too many of the characteristics of the parent faculty and does not develop its own identity. A strong, inter-faculty steering committee of staff members from both faculties (and others such as social sciences) which is responsible for programs at undergraduate and graduate level, may help to overcome bias.

In tropical countries, a single faculty of agriculture is more likely than of forestry. In this case, agroforestry can be developed within the faculty and key subject personnel who represent the absent faculty recruited.

One thing is clear: the institutional structure can facilitate the development of agroforestry, but in the final analysis it is the dedication and enthusiasm of the individual staff that will count for success. The most important factor is to have a clear idea of the goals.

References

Bene JG, Beall HW and Cote A (1977) Trees food and people: Land management in the tropics. IDRC, Ottawa

Hoskins MW (1987) Agroforestry and the social milieu. In: Steppler HA and Nair PKR (eds), Agroforestry: A decade of development, pp 191–203. ICRAF, Nairobi

Lundgren B (1982) Introduction. Agroforestry Systems 1: 3–6

Spears J (1987) Agroforestry: a development-bank perspective. In: Steppler HA and Nair PKR (eds), Agroforestry: A decade of development, pp 53–66. ICRAF, Nairobi

Steppler HA (1987) ICRAF and a decade of agroforestry development. In: Steppler HA and Nair PKR (eds), Agroforestry: A decade of development, pp 13–24. ICRAF, Nairobi

Steppler HA and Nair PKR (eds)(1987) Agroforestry: A decade of development. ICRAF, Nairobi

Zulberti E (ed)(1987) Professional education in agroforestry – Proceedings of an international workshop. ICRAF, Nairobi

Agroforestry Systems **12**: 49–56, 1990.

Keynote paper

Agroforestry training: global trends and needs

RICHARD F. FISHER
Utah State University, Logan, UT 84322-5215, USA

Abstract. A major worldwide trend toward the use of agroforestry and other sustainable agricultural systems has heightened the need for training. Such training is currently underway on every continent. This paper addresses the general principles and practices of training in agroforestry focusing on who needs training, what training is required, and designing training programs. The breadth and depth of training required by the various clientele groups – villagers, politicians, technicians, and professionals – are quite different. Politicians require broad but rather shallow training in agroforestry making them aware of the physical and biological constraints, as well as the social and economic aspects of agroforestry. Villagers require applied, hands-on training but teaching principles also enables them to develop and modify their own systems. Technicians and professionals both need more in-depth and thorough training consisting of both principles and practices. Spaid's approach to training involves the 4-D program: Define, Design, Develop and Deliver. Another model for training, the critical events model, emphasizes the need for feedback and evaluation in every stage of the training program. If a series of well-defined steps is followed, valuable, efficient, effective training programs that further the understanding and practice of agroforestry can be a reality.

Introduction

There is a major worldwide trend toward the use of agroforestry and other sustainable agricultural systems. This trend, which has been noticeable over much of the globe for nearly a decade, is now taking hold in North America, home of intensive industrial agriculture. This new importance of agroforestry has heightened the need for training. Such training is currently underway on every continent. Rather than reviewing any of these specific training programs, this paper will address general principles and practices of training.

There has been very little written about agroforestry training, although considerable training has taken place. This paper will rely heavily on conversations between the author and various trainers. I realize the danger of such an essay becoming a record of oral history and folk wisdom. I shall attempt to avoid this pitfall by addressing principles and disdaining 'war stories'.

When one considers any kind of training several central questions arise. Who needs to be trained? What kind of training is required? Where should the training be conducted? What format should be used for the training? How much training is required and how often will it be required? These questions, or rather their answers, are central to formulating effective agroforestry training programs.

Who needs training?

As we consider this question we get our first glimpse of the complexity involved in such training. Future questions will only build upon this complexity. Villagers need agroforestry training, so do politicians, technicians and professionals. Obviously the type of training required by these various groups is quite different.

Politicians require broad but rather shallow training in agroforestry. They need to be made aware of the physical and biological constraints on agroforestry, but they need not thoroughly understand the physical and biological concepts that underlie the practice of agroforestry. In addition, they must have a better understanding of the social and economic aspects of agroforestry. Trainers must keep in mind that politicians are very busy people and any training program must be well organized and targeted to address a few thoughtfully chosen points. However, many politicans have staff that work for them who can profit from more intensive training.

Villagers require applied, hands-on training. Although we tend to train such people to apply specific systems, they are capable of learning principles. Teaching principles has the advantage of enabling the trainees to develop their own systems. The level to which villagers can be trained has not been well explored. The efforts of ANAI (*Associacion Nuevo Alquimistas Internacional*) in Costa Rica indicate that with careful one-on-one instruction, villagers can become highly trained agroforestry practitioners.

Technicians require much more in-depth and thorough training. Their training must consist of both principles and practices. They must understand at least some of the theory underlying the practices they will later employ. Technician training must emphasize principles and practical methods of applying those principles to everyday problems. This means that problem analysis must be a major component of such training. The major danger in technician training is that technicians may learn a few systems very well and then attempt to apply those systems in every situation.

Technicians will most often be asked to apply specific agroforestry systems, but they must know the principles that underlie the system, and they must understand the problem that the application of the system is designed to solve. The more knowledgeable the technician is about problem definition and basic agroforestry principles, the greater the chance is that he or she can successfully apply any agroforestry system. Since agroforestry is a social and a biophysical endeavour at the same time, technicians must be broadly trained.

This same breadth of training is essential for professionals. We are currently beginning to train true agroforestry professionals, and highly specialized programs have been established to accomplish this end. However, the demand currently outstrips the supply, so that we must train existing professionals in agroforestry. This presents unique problems because the existing professionals have vastly variable backgrounds. This is an opportune point for us to consider the second major question pertaining to agroforestry training.

What training is required?

Clearly, the training required by politicians and villagers is quite narrow and fairly case-specific. The training required by technicians and professionals must be much broader and certainly is seldom case-specific.

Spaid [1986] has laid out a very sound philosophical approach to the development of training programs. He urges us to follow what he calls the 4-D program: Define, Design, Develop and Deliver. Here we are concerned with the first D: Define. What is the goal of the proposed training? Generally we hope to encourage villagers to employ a new system that is compatible with their economy and culture. Thus we can concentrate on the biological and physical components of the system. If, as is sometimes the case, the system we hope to employ requires a modification of the social or economic system, the training must be focused differently. In either case, if we properly define the problem we can design appropriate training.

Technicians and professionals generally require more comprehensive training. They need to understand not only bio-physical principles but socio-economic aspects as well. Thus the problem definition will be quite different. Commonly, technicians and professionals already have some background in biology, economics, sociology, etc., and each trainee may require training emphasis in a different area. This is the greatest challenge to agroforestry trainers. What if some trainees require a great deal of biological training and others are bored by instruction in an area they already know well?

Designing training programs

This leads us to consider Spaid's second D: Design. One solution is to group students on the basis of their strengths and proceed with narrowly focused programs. An alternative approach, and one that seems to work especially well with professionals, is to use the varied strengths of a mixed group of trainees to provide instructional support within the program. For example, this approach has worked well for the Organization of Tropical Studies.

In a mixed group, social scientists help to make biologists aware of the social aspects of agroforestry, while biologists help to make social scientists aware of the biological constraints that operate within ecosystems. If the training team also contains biological, physical and social scientists, this mixed-group format can be quite successful. These mixed groups tend to highlight the complexity of agroforestry endeavours, and remind us that our specialized training has left us with large blind spots.

There are several other important considerations in designing training programs. In what language should the training be conducted? Where should the training take place? What training format should be employed?

Obviously politicians and villagers must receive training in their native language. For technicians this is sometimes suboptimal since they most often have limited comprehension in a second language. Efforts should generally be made to provide technicians with training in their native language. Professionals have generally developed sufficient skills in a second language that they can profit equally from first or second language training. Offering training in languages such as English increases the scope of materials and the variety of personnel that can be incorporated in the training program.

Where should the training take place? It can be carried out either in the country where it is to be applied or elsewhere. It can be conducted in the classroom of the field. The location will depend largely on the defined objects of the training. There are advantages and disadvantages to any location.

In-country training has the advantage that the trainees are familiar with both the biophysical setting and the customs, food, etc. This is a powerful advantage in training villagers. The disadvantage to such training is that customs and prejudices often make the learning environment less than ideal. This is a powerful disadvantage in training professionals. However, when professional training takes place wholely in an environment different from the professional's work environment, the learning environment is also limited. Tropical training is difficult to carry out in a temperate setting.

Training can also take place either in the classroom or the field. There is a tendency for villager training to occur in the field and for professional training to occur in the classroom. Although there is little to be gained by bringing villagers into the classroom, keeping professional training out of the field is a large mistake. Agroforestry happens outdoors. Many things can best be taught in the field where phenomena can be observed first hand. In most instances, 100 percent field courses are far superior to 100 percent classroom courses and some mixture of field and classroom instruction usually achieves the best results.

What format should be used for training. Should it be lecture or laboratory, focus on case studies or principles, etc.? As with other questions regarding agroforestry training there are a number of potential answers. Villagers usually learn best when they physically participate in the instructional activity. Therefore, a field laboratory format is generally most successful with villagers. Technicians acquire skills most readily from a similar format, but usually mastering of some principles is required, and those are often best presented using a lecture-discussion format. Discussion appears to be an essential part of successful training and should be encouraged in any format.

A good deal has been written about the use of the systems/case study approach in agroforestry teaching [Huxley 1987; Bawden et al. 1984]. The strength of this approach is that students see complete packages and the problems that have arisen in the implementation of them. The major weakness is that students tend to transfer the entire system to new locations without carefully analyzing whether or not the system is appropriate for the new situation. I have found that the systems/case study approach is quite effective if the students have a thorough understanding of basic principles, and if care is taken to define the problem addressed in the case, to discuss how well the system used fits the problem, and to determine why the system used succeeded or failed. At some point either the students or the instructor must generalize from the case being studied to principles.

Developing training programs

After training objectives have been defined and a training strategy has been designed, the training package must be developed. Organization is extremely important in training. Generally we wish to pack as much material as possible into training programs so that we can maximize the benefit/cost ratio [Kearsley 1982]. In order to do this effectively, careful scheduling and organization are essential. Wasted time is wasted money. But students frequently suffer from information overload. Time for relaxation can easily be included in training programs if wasted time can be kept to a minimum.

The use of 'experts' is usually inappropriate in villager training, but experts are essential in politician, technician, and professional training packages. The problem with experts is that they present discrete viewpoints

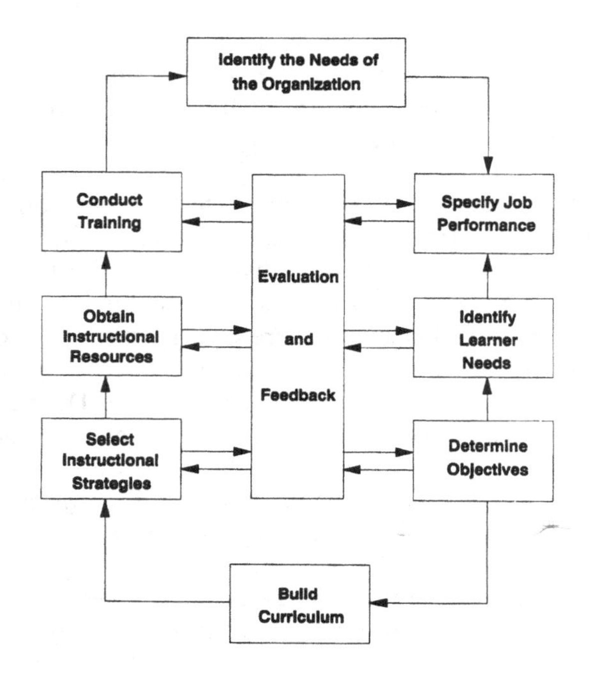

Fig. 1. Nadler's critical events model.

or packages of material and do not provide a balanced, integrated view. This can be overcome by having a trainer who ties together the material provided by the guest experts. In this way the most knowledgeable people can be used without the material becoming fragmented. Once the delivery of training begins, evaluations can be used to assess the success of the efforts at integration.

The critical events model

Nadler [1982] has provided us with a model for the design of training programs (Fig. 1). This model, the critical events model (CEM), is not too dissimilar to Spaid's 4-D approach. It does, however, emphasize the necessity for feedback and evaluation in every stage of a training program.

Obtaining adequate feedback in the various stages of a training program is not easy. It is essential during the definition or identify-the-needs-of-the-organization phase, but who knows what the needs really are? Most often people within the organization know what their needs are and can provide valuable feedback. As one moves to the design and development stages, feedback is no easier to obtain, but it is important to attempt to get adequate evaluation of the proposed program design. This can come either from the organization for whom the training is being developed or from peers. It has always struck me as odd that we readily seek peer review of our research and acknowledge that such a process improves the results, but we seldom seek similar review of our teaching or training. Certainly peer review is bound to improve the product.

It is easiest to obtain feedback or evaluation of the delivery phase of training, but such information is often the most unreliable. Training evaluation is more difficult than teaching evaluation since so many more forces are at work. Trainees are often away from home, eating strange food, and facing a more arduous schedule than normal. In short, they are stressed and often not able to unbiasedly evaluate their training until several weeks after it has occurred, when their lives have returned to normal. This does not mean that evaluation during the training program is not useful. In fact, such evaluation can often identify stressful situations so that they may be addressed before stress levels become too high.

A major problem encountered in training evaluation is that trainees often enjoy the experience but do not learn much. This can be discovered by evaluating performance after training is complete. Of course this requires the cooperation of the organization for whom the trainee works. The best training programs evaluate not only trainee satisfaction, but also the satisfaction of the agency or firm that sponsored the trainee.

Training paralysis

Moris [1981] points out the very real danger of 'training paralysis'. There are two parts to this problem. In some developing countries, all of the able land managers are involved in conducting training programs and little actual land management activity takes place. Of course, eventually there are enough well-trained people that all of them are not required to be trainers. However, the status associated with being a trainer may swell the ranks and cause a considerable time lag in getting good practitioners into the field to actually manage land.

The second part of this problem involves having the most capable people constantly being trained. There is also status associated with being selected as a trainee. Many capable people enjoy training and are constantly seeking more of it. The agency or firm that sponsors training must guard against over training. If training is to be valuable, the skills and knowledge acquired must be applied to real world problems.

The questions of how much training, how often, and who pays for training must be answered by the agency or firm that sponsors training Those who offer training must, however, aid in answering these questions Such questions should always be considered by both the sponsoring and the offering organization during the definition phase of training program development.

Conclusions

Excellent training programs can only be created if systematic models, such as the critical events model or the 4-D model, are followed. It is essential that the need for training and the desired outcomes of training be clearly defined. Efforts must then be made to design a training program that clearly meets the defined objectives. The program must then be carefully developed so that it runs smoothly and efficiently. Only in this way can the training successfully transfer the desired skills and knowledge to the trainees. Finally, the program must be delivered effectively. Throughout this entire process feedback and evaluation from sponsors, peers, trainees, and employers must be judiciously used to hone the training program. If such a series of steps is followed, valuable, efficient, effective training programs that further the understanding and practice of agroforestry can be a reality.

References

Bawden RJ, Macadam RD, Packham YJ and Valentine I (1984) Systems thinking and practices in the education of agriculturists. Agric Systems 13: 205–225

Huxley PA (1987) A combined systems, case study approach for agroforestry training. In Zulberti E (ed), Professional Education in Agroforestry, pp 122–127. International Council for Research in Agroforestry, Nairobi

Kearsley G (1982) Costs, Benefits and Productivity in Training Systems. Addison-Wesley Publishing Co. Reading, Massachusetts

Moris J (1981) Managing Induced Rural Development. International Development Institute. Bloomington, Indiana

Nadler L (1982) Designing Training Programs. Addison-Wesley Publishing Co. Reading, Massachusetts

Spaid OA (1986) The Consummate Trainer. Prentice Hall. Englewood Cliffs, New Jersey

Agroforestry Systems **12**: 57–69, 1990.

Keynote paper

Resource development for professional education and training in agroforestry

K.G. MacDICKEN and C.B. LANTICAN
Forestry/Fuelwood Research and Development Project, Winrock International, Faculty of Forestry, Kasetsart University, Bangkok, Thailand

Abstract. The development of agroforestry education and training is hampered by a shortage of information on agroforestry practices and systems and by institutional constraints which limit effective transfer of existing knowledge. Generation of knowledge through research and the effective sharing of information on agroforestry are critical to the building of a solid resource base for agroforestry education. Networks of individuals and institutions can accelerate the development of resources for agroforestry education. Primary activities of an agroforestry network would include the development and dissemination of training materials on agroforestry, curriculum development and training of teaching staff. Critical considerations for the successful establishment and operation of a network include: (1) focusing the network on a problem and identifying sufficient interest, (2) personnel requirements such as an institution with a strong commitment, and (3) other resource requirements such as funds for network meetings, publications and research.

Introduction

Knowledge, which can be transferred, is the foundation of education. Agroforestry education has been constrained by the fact that while agroforestry is an old art, it is a new science, but one which has generated some useful empirical results. This lack of scientific information is compounded by the fact that even when research is done, results are often not well disseminated or available to educational institutions.

This paper presents a discussion of the major constraints to the development of agroforestry education and training, identifies some of the resources that need to be developed and provides some suggestions as to how some of the present constraints may be alleviated. It also includes a brief presentation of networking, a cost-effective strategy to develop resources that are essential for agroforestry education and training.

Constraints to agroforestry education and training

Burley [1987a] describes some of the constraints to the development of training courses of agroforestry which also illustrate the need to build a resource base for agroforestry education and training. These constraints can be put into two broad categories:

1. Institutional constraints which limit effective transfer and use of knowledge of agroforestry practices and systems, including:
   — Lack of a tradition of teaching the subject and hence of a syllabus.
   — Lack of teaching staff with agroforestry experience and expertise.
   — Lack of relevant field facilities and agroforestry experiments.
   — Uncertain employment prospects for agroforestry graduates.
2. Lack of knowledge of agroforestry practices and systems, including:
   — Agroforestry practices and the current literature on agroforestry.
   — Objective and quantitative mechanisms for describing and modeling agroforestry systems.
   — Awareness and understanding of the social and economic considerations which farmers face.
   — Identification of relevant research topics and appropriate techniques for agroforestry research.

Each of these constraints relate to a distinct lack of transferable knowledge of agroforestry. In most cases this is information which is yet to be developed or recorded. In some cases the information already exists but is not widely available.

Resources needed for agroforestry education and training

The above constraints can be addressed through development of three types of assets: (1) physical resources, (2) teaching or training staff, and (3) training materials. Physical resources, such as land, buildings, operating funds and equipment, provide the general setting for the teaching-learning system. The faculty or trainers serve as the collectors or generators, collators, and disseminators of knowledge to the students. Training materials are the sources of information on subjects to be taught.

To identify the resources that are needed for agroforestry education and training, a logical approach is to look at the contents of the curriculum that students or trainees go through. Such contents may vary according to the level and length of training, the perception of the training needs by designers of the curriculum and other factors such as national and/or institutional policies. Although there are differences between programs in the degree of emphasis on certain disciplines or subjects, the curriculum is still a very useful basis for determining what resources are needed for the program under consideration.

A list of subjects of which agroforesters should have substantial knowledge helps identify the essence of a curriculum for determining resource needs for agroforestry education and training. The checklist in Table 1 is useful for determining the optimum combination of courses in a program. For degree programs in agroforestry, as for other disciplines, a basic core of general education courses such as biology, chemistry, mathematics and physics is necessary.

Table 1 includes a wide range of possible subjects relating to the social, economic and biophysical environment, plant and animal production systems, and land use systems including agroforestry. To an educational planner, a list of this nature is useful as a tool to:

— determine what physical resources are necessary to run a training program and take steps to obtain them;
— determine the size and fields of specialization of the staff needed to teach the subjects and prepare a staff development program to strengthen the teaching/training staff; and
— determine the topics where training materials are needed.

Strategies for resource development

Developing and implementing agroforestry education and training programs at the professional level can be both expensive and time consuming. Two means of enhancing resource development are:

1. intensive development of resources in a limited number of institutions, and
2. development of a network of institutions and individuals focused on agroforestry education and training.

The most logical places to offer agroforestry training programs are at institutions of learning where strong academic units in agriculture and forestry, and preferably in economics and the social sciences, are already in existence. Such an arrangement would easily allow the use of already existing facilities such as land and infrastructure, teaching staff and training materials. It could also promote greater interaction between the related faculties which may bring about useful curricular innovations.

While much can be said about the need for development of physical resources for agroforestry education and training at the professional level, the more urgent need appears to be in the strengthening of staff who will teach courses and in the development of training materials. This is because most physical facilities available for teaching agriculture and forestry subjects can also be used for agroforestry, while it is far more difficult to be as flexible with teaching staff.

Table 1. Checklist of subjects considered necessary for agroforestry education and training at the professional level

- *General subjects*
  - Biological sciences
    - General botany (Survey of the Plant Kingdom)
    - General zoology (Survey of the animal Kingdom)
    - Taxonomy
    - Ecology
- *Physical sciences*
  - Mathematics (algebra, trigonometry and calculus)
  - Physics
  - Elementary statistics
  - Chemistry (general and organic)
  - Geomorphology and land form
  - Soils and climate
- *Social sciences and humanities*
  - Rural sociology
  - Fundamentals of economics
  - Humanities
- *Specialized subjects*
  - Plant and animal production
  - Physiology and biochemistry
  - Genetics
  - Production techniques
  - Management (including protection)
- *Land use and land management systems*
  - Land use tenure
  - Systems analysis
  - Farming systems
  - Silvicultural systems
  - Agroforestry systems
  - Soil and water conservation
  - Project development and management
  - Policy and administration
- *Engineering and harvesting*
  - Technical drawing
  - Surveying and mapping
  - Harvesting methods
- *Socio-economics*
  - Applied sociology
  - Cultural anthropology
  - Communication techniques and extension
  - Agroforestry economics and finance
  - Community development
- *Utilization and marketing*
  - Properties and uses
  - Processing techniques
  - Marketing
- *Others*
  - Field visits
  - Electives

Ideally, agroforestry subjects at the professional level should be taught by faculty-members who have acquired advanced degrees and considerable research experience in agroforestry or closely related subjects. The present reality, however, is that there are not many faculty members who possess this qualification since graduate programs in agroforestry are a recent phenomenon and are offered only in a limited number of institutions. In the absence of agroforestry teachers with graduate degrees, existing staff in agriculture and forestry and related disciplines must be utilized to teach the subjects in the curriculum. These instructors must be provided with short training courses, preferably supplemented by study tours, to provide them with a basic understanding of the principles and practices of agroforestry. Here again the critical limitation of in-depth training materials specific to agroforestry becomes apparent, as noted in the previous example.

It is equally important to bear in mind that competence in a particular field does not necessarily imply teaching competence. To improve the quality of teaching, the faculty members should undergo a teacher's training course to improve their skills in transferring knowledge to top students or trainees. A course on this subject, 'Forestry Teachers Development Course' was developed at the UPLB College of Forestry in Los Baños, Laguna, Philippines with FAO and SIDA (Swedish International Development Agency) assistance in 1979. Although external assistance is no longer available, the college continues to offer the three-month course on a yearly basis.

A major challenge, with which the 1982 International Workshop on Professional Education in Agroforestry [Zulberti 1987] agrees, is in establishing the means for collecting, collating and disseminating information that is already available. A possible strategy to facilitate access to information needed for teaching agroforestry, is to enhance the dissemination and use of the ICRAF (International Council for Research in Agroforestry, Nairobi, Kenya) bibliographic data base which contains more than 10,000 citations in areas related to agroforestry. Increased use of this data base would be a cost-effective means of enhancing literature for searches. However, many of the citations found in this data base are not readily available in most library systems, leaving a major gap. This gap could be addressed through a network of institutions focused on agroforestry education.

Apart from being able to extract information from various sources with the aid of databases and to supplement field observations, training aids such as films, videotapes, slide sets and charts are invaluable tools for classroom instruction. This is particularly true for courses that deal with topics such as farming or cropping systems, because they are difficult to describe without the aid of appropriate graphics. The development of such teaching aids is an expensive task, which can best be addressed by a "center of excellence".

ICRAF has begun this type of development through production of posters and a basic slide show on agroforestry. However, much remains to be done to provide adequate training aids for agroforestry instruction.

One strategy that is likely to be a valuable and cost-effective means of building a resource base for agroforestry training and education, is the development of a network of individuals and institutions involved in agroforestry education and training. Brunig [1987] cites the need for international and interdisciplinary cooperation in agroforestry education. A recent international workshop on the education of forest technicians also recommended the establishment of regional and global information exchange networks of forestry technician schools [Anon. 1988]. The network approach promotes international and interdisciplinary cooperation in the development of staff, and in the generation and dissemination of information. The following sections deal with some of the critical issues which must be addressed if such a network is to be effectively developed.

Definitions of a network

Burley [1987b] defines agriculture and forestry research networks as informal or formal arrangements of cooperation between institutions with similar conditions and problems, but without the immediate resources for finding solutions to these problems individually. MacDicken et al. [1986] found three common concepts of network research among forestry and agriculture researchers in Asia:

1. Networks are loosely grouped associations of researchers who divide research problems into work assignments which are then carried out primarily by a lead institution. An ideotype of this definition is found in the approach adopted by a working group meeting organized by the International Union of Forest Research Organizations (IUFRO) as described by Shea and Carlson (undated).
2. Networks are composed of participants who conduct a set of identical or similar trials for the purpose of solving specific research problems. This approach concentrates greater resources on a given problem, and is frequently used by international agricultural research centers to test food crop varieties as described by Plucknett and Smith [1984].
3. Networks are primarily communications – oriented and may simply be groups of people who communicate in areas of common interest. Organizations which produce research publications and newsletters such as the Nitrogen Fixing Tree Association (NFTA) exemplify this type of network.

A fourth type of network might be called an integrated network which combines aspects of all three of the above concepts. This type of network would facilitate the training and development of staff and include both publications and research components. An example of this integrated type of network is the Multipurpose Tree Species Research Network which is supported by the U.S. Agency for International Development (AID), Forestry/Fuelwood Research and development (F/FRED) Project. This project is supporting network development through a series of network meetings, cooperative research projects, training, publications and small research grants.

Several other existing networks that are relevant to agroforestry education are:

1. Agroforestry Research Networks for Africa (AFRENA) based at ICRAF in Nairobi
2. The Multipurpose Tree Species Research Network based at the Faculty of Forestry, Kasetsart University in Thailand
3. The Asia-Pacific Forestry Education Network supported by FAO.

These networks are focused on topics relevant to agroforestry research and/or education, yet none deal directly with agroforestry education or the resource inputs needed for quality training programs in agroforestry. If a network approach is to be used to deal directly with the resource needs of institutions for agroforestry training and education, then a new network must be created.

Factors for success

For a network focused on agroforestry education to be successful, the organizers must carefully consider a number of factors [Plucknett and Smith 1984; Burley 1987b]. These factors must be critically examined before a decision is made to proceed with any type of serious network activity. There are many 'networks' which exist only on paper due to the lack of solid commitment on the part of donors and network participants. Certainly this is less true for networks which consist only of irregular communication links than it is for a network designed to encourage cooperative activities such as development of curricula and training materials.

Critical assumptions, which must exist before a network on agroforestry education can be effectively developed, include a clearly defined problem which is widely shared and sufficient interest and commitment to warrant further development. Table 2 outlines some of these important assumptions and the types of resources required for each stage of network development.

Table 2. Resources required for successful network establishment

Stage of network development	Critical elements	Personnel requirements	Other resource requirements
Organizational	The focus of the network (a topic or problem which is clearly defined, and is common to many countries).	Subject matter specialists who actively consult with network participants in designing network structure and activities.	Adequate funds for initial infrastructure development and travel for organizational purposes.
	Sufficient interest in network topic to warrant further development.	Initial commitment is most likely to come from an institution with strong commitment, but without assurances of future funding.	Support limited to office space and facilities for the organizer until funds can be found for the establishment stage.
Establishment	Adequate funding allocated for a minimum start-up period of five years.	Subject matter, training specialists, publications and administrative staff.	Funds for network meetings, publications and research.
	Participants willing to share both research results and training materials.	Training of teaching staff and developing of training materials.	Donor support during this phase is critical to long-term success since it is during this phase that the greatest financial inputs are required.
	Participants willing to commit resources for future network operations.		
Operational	Network sufficiently well developed to operate with reduced funding requirements.	Subject matter specialists, publications and administrative staff.	Funds for network meetings publications and research. A greater share of the required funding support for network operations should come from participating institutions.

Table 2. (Continued)

Stage of network development	Critical elements	Personnel requirements	Other resource requirements
Operational (continued)	Effective secretariat established in a "neutral" institution.	Personnel requirements during this stage should be less than those during the establishment stage due to increased direct institution to institution activity.	A major source of funds should be the regular operating budgets of participating institutions in the form of personnel support for in-country activities, publications and communications costs.
	Staff from participating institutions have sufficient training and expertise to make a significant contribution.		
	Networks guided by strong and effective leaders who have the confidence of the participants and who have recognized skills in the areas of network focus.		

Stages of network development

Network development can be classified into organizational, establishment and operational stages. Networks need strong, sustained effort over a long period of time. This requires strong leadership from a coordinating body which can initially serve as network organizer. The organizer may serve only to initiate network activities, and then pass on the coordination of network establishment and development to another institution or individual. An example is the F/FRED Project, initiated by U.S.-AID but with network development and management of network activities contracted to Winrock International which provides full-time network specialists.

In many networks the role of the network organizing body evolves from that of organizer to that of coordinator of activities during the establishment

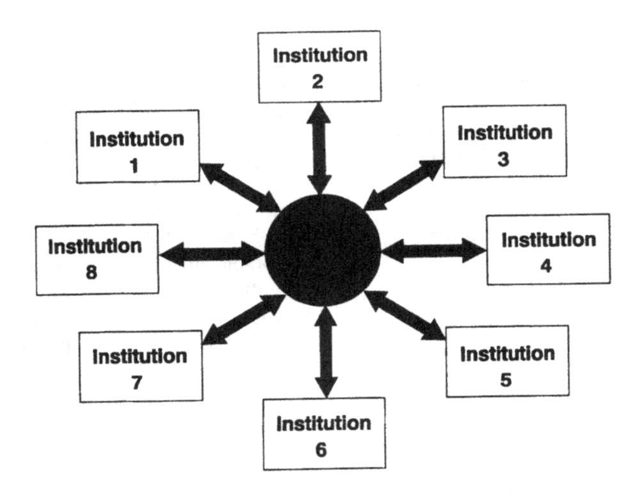

Fig. 1. Organizational stage of network development. A central coordinating body establishes working relationships with institutions and individuals who share interests in common problems.

stage. Examples of this include the Asian Cropping Systems Research Network managed by the International Rice Research Institute (IRRI) and the Asia-Pacific Forestry Education Network which was organized and is coordinated by FAO. The ultimate stage of evolution for the network coordinator is a shift to the role of a facilitator. The facilitator is able to assist with initiatives which are largely proposed and implemented by network participants.

Network building is a long-term process. Networks such as the Asian Cropping Systems Network have required strong institutional guidance from IRRI for over a decade. Another example is the F/FRED, with a goal to establish the Multipurpose Tree Species Research Network in Asia over ten years.

Organizational stage

If the elements in Table 2 exist, the next step in network organization is the identification of an institution or individual with the interest and resources to contact institutions and take the initiative to begin the organization process. This organizational stage of development requires the establishment of strong ties between the coordinating body and participating institutions (Figure 1).

Organization of a network does not necessarily require extraordinary resources other than funds and facilities for communications, including international travel. Meetings such as this workshop provide unique opportunities for potential network members to exchange ideas and make preliminary commitments to active network participation. The organization stage ends when commitments have been made and adequate funds and personnel are available to begin working on the network activities defined by the organizer and potential participants.

Establishment stage

For the network to move into the establishment stage adequate funding must be available for a minimum period of five years. Participants must be willing to commit resources and to share both research results and training or curriculum materials.

During the establishment stage, the linkages between institutions must be fostered through activities such as a newsletter, regular exchange of training materials, network meetings, curriculum development activities, staff exchange and study tours. One mechanism for establishing a network is to have a neutral coordinating body which can both administer funds and introduce a practical and functional network structure which will have a lasting role in sustaining the network.

During this stage, the coordinating body also must actively encourage and facilitate linkages between institutions while at the same time maintaining regular contact between participating institutions/individuals and the network coordinator (Figure 2). Training and institutional strengthening may be critical inputs for some participants to be able to effectively contribute to network activities.

Operational stage

The third stage of development might be a network led by a 'secretariat' or coordinating unit which can be sustained with little external resources. In this stage, major network initiatives come from network participants with the secretariat serving as a facilitator for network meetings, research grants and publications (Figure 3). The operational stage of network development requires an effective secretariat and staff from participating institutions with sufficient training to make a significant contribution. The network must be led by effective leaders who have recognized skills which are relevant to the objectives.

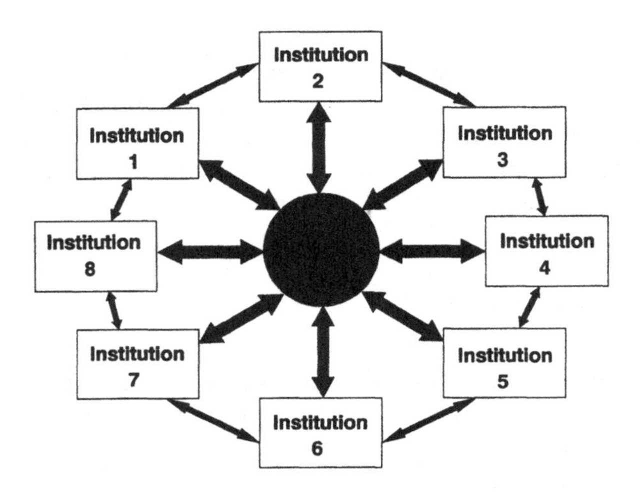

Fig. 2. Network establishment phase: Coordination/facilitation by organizing body.

Conclusion

Agroforestry education is hampered by a lack of both solid information on agroforestry land-use systems and a lack of teaching resources to effectively transfer existing information to students and trainees. Two general approaches are recommended to address these constraints. The first is to

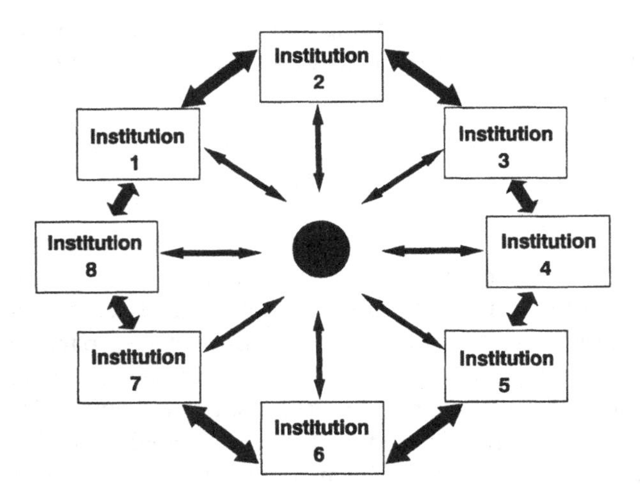

Fig. 3. Effective network operation stage: Limited facilitation role for organizing body.

intensify staff development at institutions which have both agriculture and forestry faculties or departments. The second is the formation of a network focused on agroforestry education. Primary activities of this network would include the development and dissemination of training materials on agroforestry, curriculum development and training of teaching staff.

Such a network would require commitment of financial resources for a minimum period of five years, and would require active organization efforts by an institution/individual. A number of critical assumptions must be validated before such a network development activity can be successfully initiated.

References

Anon (1988) Report of the international conference: Educating Forest Technicians into the 21st Century. Paul Smith's College, Paul Smiths, New York, August 15–22, 1988

Brunig E (1987) The network and twinning concepts of research and training. In: Zulberti E (ed). Professional Education in Agroforestry, pp 139–145. ICRAF, Nairobi

Burley J (1987a) Constraints on agroforestry teaching at the professional level. In: Zulberti E (ed). Professional Education in Agroforestry pp 151–155. ICRAF, Nairobi

Burley J (1987) International forestry research networks – objectives, problems and management. Unasylva 157/158, Vol 39

Cooper L (1984) The twinning of institutions; its use as a technical assistance delivery system. World Bank Tech Paper 21, 46 p

IUFRO (1984) Proc. IUFRO Research Planning Workshop for Asia at Kandy, Sri Lanka. IUFRO, Vienna

MacDicken KG, Dove MR, Brewbaker JL and Hyde WF (1986) Multipurpose Tree Species Networks for the Forestry/Fuelwood Research and Development Project. Winrock International Institute for Agricultural Development. Arlington, Virginia

Plucknett DL and Smith NJH (1984) Networking international agricultural research. Science 225: 989–93

Shea KR and Carlson LW (Undated) Increasing Productivity of Multipurpose Tree Species: a Blueprint for Action. IUFRO Planning Workshop for Asia, Kandy, Sri Lanka, July 16–28, 1984. USDA Forest Service. Washington, DC

Zulberti E (ed) (1987) Professional Education in Agroforestry. ICRAF, Nairobi

Agroforestry Systems **12**: 71–79, 1990.

Regional/Country report

Agroforestry education and training: an African experience

E.O. ASARE
Institute of Renewable Natural Resources, University of Science and Technology, Private Mail Bag, Kumasi, Ghana

Abstract. Agroforestry in its broadest sense has been practiced in Africa for a long time. The most common traditional form, shifting cultivation, has largely broken down due to pressures from the increasing population. Research at international centers focuses both on the development of new and the improvement of traditional systems. A main constraint is the shortage of professional and technical staff and a lack of knowledge concerning agroforestry by policy makers. Therefore, a great emphasis has been placed on education. The Institute of Renewable Natural Resources, Ghana, has begun a program in Agroforestry Education and Training. This example is used to illustrate one approach to increasing agroforestry education at African institutions.

Introduction

Agroforestry, the growing of trees and crops on the same land with or without animals, is an important land use system in Africa. As an art it has been practiced by our people for several decades. As a science it has only received serious attention in recent years. There is the need to improve present agroforestry systems and to introduce new ones. This is because shifting cultivation, the most widespread system of farming practiced by the small scale farmer who produces the bulk of the food in most African countries, has broken down due to the pressure on land from increasing populations.

Agroforestry projects are being introduced in most African countries with the assistance of national governments, donor agencies and international research organizations like the International Council for Research in Agroforestry (ICRAF), the International Institute for Tropical Agriculture (ITTA) and the International Livestock Centre for Africa (ILCA). One of the main constraints on the rapid introduction and implementation of agroforestry projects is the shortage of professional and sub-professional staff and the lack of knowledge on the part of policy makers and planners. Therefore, any step taken to assist in the provision of trained personnel will go a long way in assisting in promoting agroforestry practices.

Introduction of agroforestry courses in institutions of higher learning in Africa has taken place only recently. We, therefore, stand to benefit greatly from sharing views and consulting with participants from all over the world. For example, we would like to share our experience in the introduction of courses in Agroforestry Education and Training at the Institute of Renewable Natural Resource (IRNR), Ghana.

The IRNR was established in 1982 as a Faculty of the University of Science and Technology (UST), to promote through teaching, research and extension, the proper management and utilization of the forests, savannas, wildlife, freshwater fisheries and watersheds of Ghana. Three of the courses offered at the institute, a three year B.Sc., a two year undergraduate course in natural resources management and a one-year post-graduate course in agroforestry are relevant to this discussion.

The first two courses offer four options: forestry, wood science, wildlife and range management, and freshwater fisheries. These options are designed to produce professional and sub-professional personnel with a multiple-use management approach to the development and utilization of natural resources. We believe that students must be aware of the interdependence of phenomena and that a training program based on the ecosystem concept will enable our graduates to plan for the whole rather than any one particular resource. They will also be able to discuss intelligently with and seek advice from other specialists.

Background to the mounting of agroforestry courses

Ghana is an English-speaking West African nation occupying an area of 240,000 km^2. It has a population of about 13 million people, which may reach 20 million in the year 2000. Ghana's economy is mainly based on agriculture. About 70% of the people are farmers. Eighty to 90% of the food is produced by a small-scale farmers. The most important export commodity, cocoa, contributes 73% of the foreign exchange earnings and is produced almost entirely by small farmers. It is estimated that 8.23 million hectares of Ghana's forests are being destroyed at the annual rate of about 0.39%; while the 15.63 million hectares of savanna woodland are being removed at the annual rate of 0.51%. The present system of shifting cultivation has broken down in several areas due to the shortening of the fallow period from 7–15 years to 1–3 years. This, and the relatively fast rate of deforestation and 'desavannisation', have given rise in many areas to a lowering of soil fertility resulting in reducing crop yield, creation of desert-like conditions especially in the semi-arid and sub-humid zones, and a scarcity of fuelwood giving rise to high fuelwood prices. It is against this background that the government of Ghana in 1984 decided to introduce agroforestry practices throughout the country to assist in reversing the rapid rate of depletion of the forests and savannas and increasing agricultural production and maintain ecosystems.

Introduction of agroforestry course

In order to implement this policy, the government established an Agroforestry Unit at the Ministry of Agriculture. It also appointed a National Agroforestry Committee to formulate policies on agroforestry, advise the government on agroforestry matters and co-ordinate agroforestry projects being undertaken by both governmental and non-governmental agencies.

The government requested FAO to assist in the establishment of the Agroforestry Unit and to provide a consultant to determine the immediate needs of the unit and make recommendations for assistance in the implementation of agroforestry programs. At the same time, the government requested the IRNR to train the personnel needed for the implementation of agroforestry projects. It was in response to this call that the IRNR set out to introduce courses in agroforestry. In the planning stage, our objectives were:

1. To incorporate agroforestry in existing courses to give students a broad knowledge.
2. To introduce a three unit agroforestry course in the final year of the B.Sc. and Diploma Courses and to encourage students to select agroforestry topics for their final year through field projects.
3. To introduce an agroforestry option at the B.Sc. and Diploma levels to produce generalists who can identify and tackle problems with the help of specialists wherever necessary.
4. To mount a one-year post-graduate diploma course in agroforestry to train professionals mainly for extension work.
5. To establish a two-year M.Sc. course and subsequently a three-year Ph.D. course to train specialist agroforesters in research and education.

Action taken

It was felt that we were not in a position initially to introduce an agroforestry option, or M.Sc. or Ph.D. courses, as the establishment of these would overstretch our resources. We didn't have adequate resources, such as staff, laboratory and field facilities, text books and readily accessible literature, teaching aids, etc.

We were, however, in a position to accomplish the first three objectives. Although in most of our courses students were made aware of the need for agroforestry and the advantages of multiple land-use, we decided to look again and identify areas where there was a need to increase the agroforestry content. Thus we decided to introduce immediately a three-unit agroforestry course in the final year of both the B.Sc. and Diploma (under-graduate) courses as this was within our resources and competence. These measures, however, will hardly produce agroforesters.

Agroforestry diploma course

In order to meet the immediate and greatest need of the country, we decided to mount a one year post-graduate diploma course in agroforestry to train extension workers. During the initial stages, students would be drawn mainly from professional staff serving in the Departments of Crop Services and Animal Health and Husbandry of the Ministry of Agriculture and the Department of Forestry of the Ministry of Lands and Natural Resources. A great deal of emphasis would be placed on practical training and field work as the professionals would be working in the field with farmers.

There are two types of courses, core courses to be taken by all students and optional courses from which students may be exempted depending on their background (Appendix 1). For instance, a student with a degree in forestry has to take all the optional courses except Fundamentals of Forestry for Agroforesters, while those with a degree in agriculture must take Principles of Range, Wildlife and Freshwater Fisheries Management. Group field projects, field visits, presentation of seminars and literature research will be featured.

Each student will be required to produce a thesis based on field studies. Specialists from governmental and non-governmental agencies as well as scientists from local and international research organizations will be invited to give lectures, conduct seminars, workshops, run short courses, etc. The Diploma Course is being handled by our Agroforestry Unit, which is to be turned into a department at the appropriate time.

Steps taken

It was envisioned that it would take two to three years to complete the initial program initiation. The first step taken was the recruitment of staff. As there were no agroforesters, graduates in agriculture and natural resource management were recruited and sent overseas for training in agroforestry at CATIE in Costa Rica and Britain. One of them with a Ph.D. participated in the ICRAF/USAID Training Course on Research for Development in 1987. One expatriate with an M.Sc. in forestry and its relation to land use from Oxford, who had field experience in agroforestry and social forestry in Sri Lanka, was recruited.

Steps were also taken to provide facilities for field studies. Alley farming plots were established using *Leucaena leucocephala, Gliricidia sepium, Ficus exasperata, Baphia nitida* and *Pithecolobium dulce*. An orchard of multipurpose trees and shrubs was started where both local and exotic species are being introduced.

The question of the preparation of curricula, course description and syllabi was then tackled. In this regard, the director spent one week in 1986 and one month in 1987 at ICRAF identifying and collecting teaching materials, literature and teaching aids.

Agroforestry training

In addition to agroforestry education, the IRNR has also been involved in agroforestry training. The Ministry of Agriculture, which has embarked on the establishment of alley farming demonstration plots in the 10 regions of Ghana, requested in 1987 the institute to run short courses in agroforestry with emphasis on alley farming for technicians in charge of agroforestry demonstration plots. Three courses have been held thus far for about 45 technicians drawn from the Departments of Crop Services, Animal Health and Husbandry and some non-governmental organizations. The courses last three to five days each. The following topics are treated:

— Introduction to agroforestry
— Alley farming system
— Rural forestry
— Establishment and management of forest trees and nurseries.

In each case slides and films illustrate the lectures. Participants also visit the alley farming and multipurpose tree crop plants at the institute's farm. Short talks by students on agroforestry activities in their districts and group discussions form an important part of the program.

Future plans

In retrospect, whatever gains we have made so far are due to a number of favorable conditions. The government of Ghana was keen in promoting agroforestry and took appropriate measures to ensure successful implementation of agroforestry projects. The responsible government departments were committed to the implementation of government agroforestry policies and were conscious of the fact that agroforestry projects could not have been implemented without trained manpower. IRNR was in place, capable and ready to mount courses in agroforestry education and training. International agencies were ready to assist. For instance the U.N. university provided the seed money for the establishment of alley farming plots, and sponsored the training course in agroforestry of one of our lecturers. FAO provided funds for the first agroforestry training course; ICRAF assisted in the development of the courses and supplied literature and teaching aids. The ODA (U.K.) sponsored training in agroforestry for three of our lecturers.

The continued assistance of the Ghana government and international research organizations as well as cooperation of government departments is necessary if we are to succeed in producing the right calibre and type of trained personnel. We plan to intensify and expand our training program to cover not only technicians but also policy makers and planners. The goal of the institute is to strengthen its research capability and enhance its teaching by participating in research activities of AFNETA (Alley Farming Network for Tropical Africa) and to also collaborate with ICRAF in agroforestry research in Ghana. We believe that with the passage of time we shall gather enough experience, acquire the necessary facilities as well as have the full complement of staff to enable us to embark on an M.Sc. and later Ph.D. program in agroforestry.

Appendix 1

Institute of Renewable Natural Resources, University of Science and Technology, Kumasi, Postgraduate Diploma Course in Agroforestry

In 1985, the government of Ghana directed that a Department of Agroforestry should be created within the Ministry of Agriculture to promote the implementation of agroforestry projects throughout the country.

The course is at the postgraduate diploma level. This decision was arrived at after consultations with Nairobi's International Council for Research in Agroforestry (ICRAF) and other organizations as well as a study of undergraduate and postgraduate courses in agroforestry run by several other institutions.

The course was organized by the Agroforestry Unit of the Institute, a unit that will become a department at the appropriate time. The course is taught by the staff of the IRNR. Whenever necessary, staff from other faculties of the University will be called upon to assist. Short-term visiting lecturers/professors, sponsored by donor agencies, also teach special courses and give seminars.

The course is open to holders of B.Sc. in Natural Resources Management, Agriculture,

Biological Sciences. Graduates in other disciplines with relevant field experience may also be considered for admission.

Agroforestry research and demonstration plots established at the institute farm located on the university campus are used for teaching and demonstration purposes. Field trips are taken to different parts of the country to familiarize students with agroforestry practices and problems.

The course duration is two semesters. It terminates at the end of the second semester but students undertake a four-week field experience during the long vacation. A great deal of emphasis is placed on practical/field training which involves both individual and group projects. Students are required to pass all core courses and to take at least one optional course each semester.

The course structure is indicated below:

First semester				
Course no.	*Core courses*	*Lecture*	*Practical*	*Credit*
AF.501	Concepts and practices of agroforestry I	3	3	4
AF.503	Land evaluation and classification I	2	2	2
AF.507	Seminar	–	–	–
AF.509	Group projects	–	–	3
Optional courses				
AF.511	Fundamentals of crop science	2	2	2
AF.513	Fundamentals of forestry for agroforesters	2	2	2
AF.515	Principles of range, wildlife and freshwater fisheries management	2	2	2
Second semester				
AF.502	Concepts and practices of agroforestry II	3	3	4
AF.504	Land evaluation and classification II	2	2	2
AF.506	Agroforestry research techniques	2	3	2
AF.508	Seminars	–	–	1
AF.510	Group projects	–	–	3
AF.512	Thesis	–	–	4
Optional courses				
AF.514	Fundamentals of animal science	2	2	2
AF.516	Extension methodology	2	2	2

Syllabi for courses

1. AF.501: *Concepts and practices of agroforestry I*

   1. Global aspects of agroforestry land-use: Systems, classification of agroforestry; soils/

climates and ecozones and choice of systems, components and management practice; definitions.

2. Technical/Scientific concepts (Introduction); Plant component, with special reference to multipurpose trees; factors affecting biomass production; soil sustainability; site enrichment processes; environmental resource sharing; plant management in AF; shelter effects; etc.

2. AF.502: *Concepts and practices of agroforestry II*
   1. Data sources on land use: climate, soils, production, etc.
      i) in Ghana
      ii) relevant world-wide: e.g., Spot, Landsat, FAO agro-ecological zones etc.
   2. Policy issues: general policy planning, policy relevant to grassroot development and AF technologies.
   3. Clustering/Aggregation/Selection techniques for landuse systems in Ghana: land use classification systems, identify examples of some land use systems in Ghana.
   4. Identification of appropriate landuse systems: introduction of land evaluation methodology, identification of successful land use systems in Ghana. Land use systems in experimental stages and potentially successful.
   5. Identification of systems chosen for land use systems in Ghana – problem and potentials.
   6. Prioritization of problems solvable by AF – preliminary 'matching' to potential solutions; introduction of Diagnosis and Design (D and D); use of D and D for problem solving.
4. AF.504: *Land evaluation and classification II*
   1. Selection procedures for choosing diagnostic field sites;
   2. Field diagnosis exercises;
   3. Elaboration of 'systems' specification for candidate technologies;
   4. Design initiation; listing/prioritizing potential interventions; listing/prioritizing general technology specifications; deriving and evaluating specific technology recommendations; outlining necessary research/extension plans.
5. AF.506 *Agroforestry research techniques*
   1. Space and time related considerations.
   2. Units of measurement and yield considerations.
   3. Simplifying AF research; logical research stages; defining research objectives; basic research approaches; MPT selection/testing; zonal AF experimentation; mixed AF experimentation; rotational plots; special investigations; introduction to on-farm trials.
   4. Agroforestry experiments; design (field layouts); analysis of data.
6. AF.513: *Fundamentals of forestry for agroforesters*
   1. Concepts and scope of forestry; significance of forest resources in rural development; forest resources administration and control.
   2. Concepts and practice of silviculture; silviculture of man-made forests; planting stock production and nursery technology; planting and seeding; tending of short-rotation tree crops; regeneration of short-rotation tree crops.
   3. Forest/woodland management; tree and forest measurements (measurement of tree dimensions; instruments and methods; inventory of forest stands and crops); forest regulation (rotation; increment, sustension of production and yield).
   4. Processing; harvesting of short-rotation tree crops; fuelwood and charcoal.
   5. Laws affecting forest trees in Ghana.
7. AF.515: *Principles of range, wildlife and fisheries management*
   1. The ecology, utilization and management of the grasslands of Africa.
   2. Grassland plants and their role in agroforestry.
   3. The important wildlife of Africa; factors affecting their distribution, abundance, etc.

4. Food and food habits of wildlife.
5. Wildlife ranching and domestication.
6. Endangered species.
7. Freshwater fishes of Africa.
8. Principles of aquaculture.
9. Design and construction of fish ponds.

8. AF.511: *Fundamentals of crop science*
1. The science and art of crop production.
2. Environmental factors in crop production.
3. Physiological aspects of crop yield.
4. Principles of crop improvement; crop breeding.
5. Cropping systems.
6. Soil conservation, soil tillage, planting and harvesting of crops.
7. Soil fertility, fertilizer application and manuring.
8. Crop protection, understanding pests (insects, fungi and weeds); control management.
9. Storage, processing and marketing of crops.

9. AF.514: *Fundamentals of animal science*
1. Cattle sheep and goat breeding and production.
2. Poultry and pig breeding and production.
3. Diseases and principles of animal health.
4. Nutrition: feed sources, quality, types, balanced diets, etc.
5. Mixed farming: livestock and crop production systems.
6. Animal products – processing, storage and marketing.

10. AF.516: *Extension methodology*
1. Principles and philosophy of extension methodology.
2. Review of extension methodology.
3. Extension education, planning and programming.
4. Sociological basis of extension.
5. Fundamental concepts of rural sociology, rural society and organization.
6. Extension service and its operation in developed and developing countries.
7. Modern trends in extension education and methodology.

Agroforestry Systems **12**: 81–86, 1990.

Regional/Country report

Education in agroforestry at the University of Melbourne

ROGER SANDS
School of Agriculture & Forestry, The University of Melbourne, Parkville, Victoria 3052, Australia

Abstract. Graduate training in agroforestry is offered at two institutions in Australia, the Australian National University and the University of Melbourne. Agroforestry at Melbourne benefits from being in a combined School of Agriculture and Forestry. Melbourne recently also commenced an undergraduate program. Agroforestry is interpreted rather broadly to include any uses of trees in the rural landscape. Most emphasis at the undergraduate level is on temperate Australia, while that at the post-graduate level is much more tropically oriented.

Introduction

In Australia, there are eight universities that offer undergraduate and graduate education in Agricultural Science and two in Forest Science. The Australian National University (Department of Forestry) and the University of Melbourne offer M.Sc. and Ph.D. degrees for research in agroforestry. The disciplines of agriculture and forestry come together in a School of Agriculture and Forestry at the University of Melbourne, where formal course work at undergraduate and post-graduate diploma levels are offered.

Even so, agroforestry is relatively new at the University of Melbourne. Research in agroforestry commenced in 1983. Post-graduate research degrees (Masters and Ph.D.) in agroforestry have been offered only since 1985. Agroforestry was offered at the post-graduate diploma level for the first time in 1988. Agroforestry will be offered as an undergraduate subject for the first time in 1989.

Undergraduate program

The University of Melbourne offers four year B.Sc. courses in Agricultural Science and in Forest Science. Each of these courses has fixed subjects and no choice is available until the fourth year of the course. Students are permitted to choose from a range of subjects during the fourth year of the course, and it is at this stage that students in either agriculture or forestry can choose some degree of specialization in agroforestry. The subject agroforestry is offered in the second semester of the fourth year of the course. A copy of the syllabus is attached as Appendix 1.

Agroforestry has been defined as growing trees and crops or animals together on the same land. However, we define agroforestry in a broader sense, as any use of trees in the rural landscape. The subject deals with the use of trees on farms in an integrated manner for providing timber, fodder, fuel and other tree products; for providing shelter for stock and crops; for conserving wildlife on farms; for beautifying the landscape; and for arresting soil degradation and rehabilitating saline areas. All of these are treated within the context of whole farm planning and land management.

The subject at the undergraduate level deals essentially with agroforestry in a temperate region with particular reference to Australia. It also aims to examine agroforestry in the world context, and to deal with agroforestry in the tropics so that graduates can seek employment in these areas if they wish.

Students undertaking the degree in Agricultural Science who elect to do agroforestry in the fourth year of course work are given the option of taking a subject called Forestry Systems in the first semester of the fourth year. The aim of this is to give these students the necessary background in Forestry to enable them to get the most out of their second semester course in agroforestry. Likewise, students undertaking the degree in Forest Science can take the subject Agricultural Systems for the same reason. Students in both Agricultural Science and Forest Science also can elect to do a project in Agroforestry during the fourth year of their course. This is a small research project which is usually coordinated with one of the larger agroforestry research programs within the School.

Post-graduate diploma

A post-graduate diploma in Agricultural Science and in Forest Science is offered by the school. This is a one-year diploma by course work which is offered to overseas students as well as Australian residents. Students can specialize in agroforestry in the post-graduate diploma. The post-graduate diploma is well supported by overseas students and it has been specifically designed to meet their needs. In order to be enrolled for the post-graduate diploma, students must have achieved a high grade in a university course in Forest Science or Agricultural Science or in some related field of Biology or Land Management. The recommended subjects for students enrolled in the post-graduate diploma and specializing in agroforestry are: Agroforestry, Project in Agroforestry, Agricultural Systems, Forestry Systems, Rural

Development, Field Studies, and Statistics and Experimental Design. For those students who transfer to a research degree, the project can be extended to become the basis of this degree. The project can be undertaken in the applicant's home country. Some of the subject matter contained in the subjects Agroforestry, Forestry Systems and Agricultural Systems is common to that provided for these subjects in the under-graduate program offered to Australian students. However, a strong emphasis has been placed on agroforestry, forestry and agriculture in the tropics in the post-graduate diploma program.

The main financial sponsorship for students to the post-graduate diploma program is from the Australian foreign aid vote under the auspices of AIDAB, although other agencies such as FAO and the World Bank also sponsor students to the program.

Post-graduate research degrees

A 2-year Master's degree and a 3-year (minimum) Ph.D. for research in an aspect of agroforestry are offered. The research projects may be located overseas or in Australia. Aid agencies again support overseas students in these programs. Post-graduate research degrees in agroforestry are also offered by the Department of Forestry, at the Australian National University, Canberra.

Conclusion

The School of Agriculture and Forestry at the University of Melbourne is committed to providing high quality professional education in agroforestry at both the undergraduate and post-graduate level (Appendix II). The intention is to provide agroforestry specialists for Australia as well as servicing the needs of other countries, mainly in the tropics. The School is essentially at the beginning of these endeavours and it welcomes the opportunity to join in this international workshop to share our experience and to learn from others.

Appendix 1

Agroforestry syllabus at the University of Melbourne

A. Trees in the rural landscape
  1. Farm trees in context
    — Trees and farm ecosystems
    — History of Australian land use and attitudes to trees
    — Emerging awareness of roles for farm trees:
      Environment, production, complementarity
    — Trees and sustainability
  2. Trees and the whole farm plan
    — Land classing and land capability
    — Integration of trees with farm management
    — Planning tree establishment and protection
    — Economic feasibility of a whole farm approach
  3. Land degradation control
    — Erosion
    — Salinity
    — Tree decline
    — Assessing needs-immediate and long term
    — Integrating trees with other control measures
  4. Shade and shelter
    — Animal comfort zones
    — Principles of climate modification
    — Assessing shelter requirements
    — "Day in day" out shelter
    — Emergency shelter in extreme conditions
    — Sheltering crops
    — Integration of shelter into the farm plan
  5. Diversification of farm incomes
    — Firewood
    — Fodder
    — Farm timbers
    — Assessing opportunities and needs
    — Site requirements
    — Management
    — Marketing
    — Integration into the farm plan
  6. Asthetics and wildlife habit
    — The farm in the rural landscape
    — Principles of landscape - can they work on commercial farms?
    — Importance of wildlife habitat - biodiversity
    — Diversity and resilience in agro-ecosystems
    — Potential for reducing inputs through bio-control
    — Patch and corridor matrix-island biogeography
    — Watercourses and wetlands
B. Agroforestry systems
  1. Agroforestry: definitions and classifications
    — Development of the agroforestry concept

- — Traditional agroforestry systems
- — Development of alternative systems

2. Tropical agroforestry
   - — Tropical forestry and land use background
   - — Development of solutions
   - — Traditional values introduced
   - — Case studies: Asia, Africa
3. Australia and New Zealand
   - — *Pinus radiata* agroforestry development
   - — Other timber species: eucalypts, poplars, black walnut, etc.
   - — Other systems: fruit, nut, seed and fodder species
   - — Alternative agricultural components

C. Processes
1. Principles of competition
2. Competition between trees and herbaceous species for water, nutrients, and light
3. Design of Agroforestry experiments
4. Simulation modelling of agroforestry systems

D. Management of agroforestry systems
1. Choice of system
   - — Manager's objectives
   - — Agricultural component
   - — Forestry component
   - — Management options
   - — Integration into a whole farm plan
2. Tree establishment and maintenance
   - — Planting
     - — Choice of planting stock
     - — Site preparation
     - — Weed control
     - — Planting method
     - — Fertilization and post planting care
3. Tree management
   - — Pruning
   - — Thinning
4. Agricultural management
   - — Pre-planting
   - — Early years
   - — Tree protection options
   - — Intercropping systems
   - — Silvicultural management phase
     - — Pasture management
     - — Stock management
     - — Shade tolerant crops and fodder

E. Economics of agroforestry
1. Comparison of agroforestry with alternate lands uses
2. Value of shelter, land rehabilitation, wildlife conservation and aesthetics as well as timber
3. Taxation and financial management
4. Harvesting and marketing of forestry products from farms

F. Extension of agroforestry
   1. Aim of an extension program
   2. Agroforestry as the land use innovation
   3. Farmers as the potential adopters
   4. Optimal extension program
   5. Case studies
   6. Tree planting groups
   7. Community involvement

Agroforestry Systems **12**: 87–89, 1990.

Regional/Country report

A brief account of professional education and training in agroforestry in China

WANG SHIJI
Research Institute of Forestry, Chinese Academy of Forestry, Beijing, Peoples Republic of China

Abstract. China's interest in agroforestry derives from the fact that less than 12% of the total land area is forested, while demand for timber and fuel is greatly increasing. Afforestation is very active, much of it in agroforestry systems. However, there are no formal programs or courses yet in agroforestry. Currently, though, it is a popular subject for graduate student research, and there are courses under other names ('ecological forestry' or 'three dimensional forestry') that do cover much of the subject. Three-to-ten day intensive training programs are more common, including lectures and visits to demonstration farm areas.

Introduction

China is a country without many forests. The forest coverage is only 12% of the total land area. As the national economy develops, the demand for timber is increasing; but so is the shortage of supply. The volume stocking in the northeastern forest areas is decreasing annually. In the heavily populated plains area, however, the extraordinary shortage of timber and the deterioration of the agricultural environment are encouraging the development of agroforestry. Because of this, the forst coverage has increased 8% during the last 40 years.

Characteristics of agroforestry in China

Agroforestry in China can be catalogued into the following groups:

1. 'Four-sided' planting: Since the 1950s, the government has devoted much effort in the promotion of 'four-sided' planting. This method usually follows along the contour of villages ('village-sided'), along the roads ('road-sided'), along the ditches ('ditch-sided'), and along the rivers ('river-sided'). Multipurpose trees are mainly used to supply the demand for civil building materials, fodder and fuelwood. They also provide shelter and aesthetic benefits. About 4.9 billion trees are estimated to have been planted in four-sided plantings, but the statistics are incomplete.
2. Intercropping: This method of farming has been in practice for more than

2000 years. The most successful trees are jujube (*Zizyphus jujuba*), paulownia (*Paulownia* spp.) and some fruit trees. They are intercropped with agricultural crops in certain spacings. For example in the North China plains, 3.18 million hectares are intercropped. This accounts for 57.7% of the total land suitable for agroforestry intercropping. Area under intercropping has increased as much during the past three years as during the previous 30 years.

3. Network of windbreaks: In the last 40 years, field windbreaks have been planted in the regions with serious wind erosion. Later, they were developed into a more beneficial type of windbreak system. Presently, the sheltered-field areas reach 10.67 million ha, accounting for 45.7% of the total arable land. During the past three years, the areas of newly-planted windbreaks amount to one third of the total areas planted in the previous 30 years.
4. Traditional types of systems: In the subtropical areas in the South, some traditional types of agroforestry systems have developed. Included are mulberry-fishpond and sugarcane-fishpond systems.
5. Watershed management areas: In the hilly areas, incorporating agroforestry with small watershed management has resulted in trees, shrubs and land covers being planted to control water runoff and soil erosion. In some areas with improved conditions, fruit and nut trees are planted following the contour of crop fields. This also provides fuelwood to local inhibitants.
6. Arid and semi-arid regions: In the arid and semi-arid regions in the Northwest, agroforestry systems are established to supply fodder, shelter and weaving materials for baskets and other products.

Education in agroforestry

Professional forestry training in China includes higher forestry universities or colleges and intermediate forestry schools. Higher forestry schools attract high-school graduates and intermediate forestry schools attract junior or middle-school graduates. The lengths of schooling are four and three years, respectively. Non-professional forestry education includes amateur correspondence and TV school. Presently, there are eleven higher forestry universities and colleges which include 85 specializations. Total on-school undergraduate students are about 12,000: graduate students (including Master's degree and Ph.D.) number about 400 and teachers about 3,800. There are 36 intermediate forestry schools, including 81 specializations. These have 1500 on-school students and 2000 teachers.

A course named agroforestry has not yet been set up in China. In China, a more popular name than agroforestry is 'ecological forestry' or 'three dimensional forestry'. The concept of agroforestry is, however, included in various specializations and courses. For example, a course 'Multi-management in Forest Areas' was set up in the Department of Economic Forestry in Mid-South Forestry College. This course is offered to junior students. It consists of 60 academic hours, which includes 40 lecture hours and 20 training hours. The main goal is to introduce the administration and cultural techniques of multi-management in forested areas. This includes intercropping methods of fruit trees, vegetables and medicinal herbs, and fresh water feeding techniques. In Beijing Forestry University, a course of soil and water conservation is given which introduces the principles of windbreak establishment and small watershed management methods using trees, shrubs and grasses. Systematic and complete courses in agroforestry at the Master's and Ph.D. levels are not available. However, agroforestry is one of the top research subjects in forestry at various universities and colleges. The reports on agroforestry research can often be seen in various journals of forestry schools.

Training in agroforestry

Agroforestry training classes are held irregularly in forest research and teaching institutes and forest bureaus. In these classes, combined teaching methods using lectures (by invited specialists) and site observations are used. The lecture time may be 3–10 days. For example, there are many peasant participants in the training classes on growing mushroom, ginseng, edible shoot of bamboo under trees, etc. Shenyang Agricultural college in Liaoning Province has held several training classes in 'garden forestry', which gives technical advice to farmers to help them increase their income.

Supported by the International Development Research Centre (IDRC) of Canada, the Chinese Academy of Forestry's Research Institute of Forestry sponsored an International Farm Forestry Training Course in May 1986 which included 27 foreign participants from ten countries. After a week of lectures in Beijing, the participants were taken to Shandong, Henan, Anhui and Zhejiang, etc., to visit paulownia-crop, jujube-crop intercropping and models of agroforestry. Personal visits to representative sites like this can teach the ecological benefits and economic profits of agroforestry. In China, one-to-two day visits to agroforestry field areas are also used as special training and several thousand people at various times can utilize these sites.

Agroforestry Systems **12**: 91–96, 1990.

Regional/Country report

Agroforestry education and training in European institutions

H.J. VON MAYDELL
Institute of World Forestry, Leuschnerstr. 91, D-2050, Hamburg, Fed. Rep. of Germany

Abstract. Integrated land uses, many now referred to as agroforestry, have a long history in Europe. In the past, the main trend was the movement of agricultural and pastoral activities into forests. The introduction of trees into non-forested (or once forested) lands is a much more recent occurrence, particularly the cultivation of high value fruit and nut trees in the Mediterranean countries, hedgerows/windbreaks in northwestern Europe and windbreaks in eastern Europe and the southern Soviet Union. Environmental concerns of intensive agriculture are increasing the demand for alternate production systems such as agroforestry. Education and training in agroforestry in Europe is very diverse because of the country specific issues regarding land use.

Introduction

Agroforestry dates back to very early times of land use in almost all of Europe. 'Integrated land-use', as it would be called today, was the common approach to utilizing natural resources. The main direction of development was to introduce agricultural and/or pastoral practices into the forest. The trend was enforced with increases in population, technical progress and exclusive use of lands for either agriculture or animal husbandry. The encroachment of forest lands continued until industrialization which started at different times in different areas and in some areas just recently. In Central Europe, for instance, the prime roles of forests once were the provision of oak and beech mast (acorns) for wildlife, cattle grazing, collection of forest litter, and edible products, hunting, etc. Timber production as the main objective started only about 200 years ago, and the importance of other forest benefits, such as recreation, environmental improvement, etc., emerged only a few decades ago. In some areas, such as the Mediterranean zone, northern Scandinavia and in most mountain ranges, use of forests for range management is still of high importance.

More recently, the cultivation of trees outside forest land on fields and pastures has gained importance. The outstanding example is the planting of poplars and other tree species like chestnut, olives, figs, etc., in the Mediterranean countries, hedgerows/windbreaks in northwestern Europe, and as forest-windbreak belts in eastern Europe, especially the Ukraine and southern Russia.

Thus, agroforestry, in its broader sense, is still widespread in many parts of Europe, but it is often overlooked or not considered an optimal land use by policy makers. This may change. Severe problems of excess food production in the EEC countries calls for alternative uses of farmlands. This is supported by environmental concerns, for which agroforestry may well offer feasible options.

Agroforestry education: site specific and demand-oriented

Agroforestry development in Europe will have to be site-specific. There are obvious differences and disparities between regions and even within small areas, for both natural and socio-economic reasons. Practices will range from reindeer management in Scandinavia to cattle grazing in the macchia from the dehesa-system in Spain and Portugal to the alm-management in the Alps, with a wide range of transition forms in between.

The same applies to tree-crop interfaces. Whereas silvopastoral systems obviously prevail on marginal sites, agrosilvicultural systems may be applicable on the best soils and in intensively managed areas with close links to horticultural systems and to agro-industries. Agroforestry may even be an option for solving socio-economic problems, such as labor shortages in specific rural areas.

There is an obvious demand to reduce excess agricultural production through land uses which are economically and environmentally acceptable. There is a demand to recultivate marginal or degraded lands, and to diversify production. There also is a demand to reduce agro-chemical pollution.

It cannot be denied that there are many constraints which impede agroforestry in Europe, and there are good reasons to keep other highly-productive current land use systems. However, agroforestry does already exist, and agroforestry offers prospects for the future and needs to be discussed. In addition, European countries have declared their responsibility and readiness to assist countries in the tropics and subtropics in rural development. Agroforestry has been identified as one option for overcoming many of the prevailing problems and constraints, and is internationally accepted as an approach to future-oriented land use. There is an increasing demand for know-how in agroforestry, and Europe is expected to accept the challenge and to contribute to worldwide efforts.

Review of ongoing programs

At a first glance, however, the institutional scene appears gloomy. Everybody talks about agroforestry, but very few universities and other educational institutions teach agroforestry. One of them is the Coleg Prifysgol Gogledd Cymru, University College of North Wales in Bangor. A recent announcement from the department [Roche 1988] characterized the program as follows:

'At the present time the department has over 80 postgraduate and undergraduate students from 28 different countries of the tropics and sub-tropics. These students follow a range of courses which the department has developed with the specific objective of meeting the needs of temperate and tropical nations for expertise in the fields of forestry, agroforestry and wood technology'. The following degrees are offered: B.Sc. Forestry, 3 year undergraduate degree program; B.Sc. Agroforestry, 3 year undergraduate degree program; M.Sc. Environmental Forestry, 1 year taught-postgraduate degree program; M.Sc. Forest Industries Technology, 1 year taught-postgraduate degree program; M.Sc. by research, and Ph.D. by research.

In relation to Bangor's agroforestry program, Roche [1988] notes that agriculture and forestry have quite separate administrations, systems of research and education and advisory services. Yet both are concerned with biology and land management. The principles underlying plant breeding and the use of fertilizers and weed control apply equally to trees and potatoes. There are few economic, ecological or social reasons for such rigid separation of these two disciplines; the reasons lie elsewhere, and have more to do with the accidents of history than land-use imperatives.

In recent years, the adverse effects of this separation have become increasingly apparent. Farmers have little or no incentive or encouragement to manage woodlands, hedgerows and trees. Foresters are uncomprehending and often unconcerned in regard to most aspects of the rural economy, other than that of monocultures of timber species for industrial purposes.

There are many adverse consequences of this. Those that manifest themselves most prominently are the undermining of the ecological, and often economic, bases of peasant agriculture in the tropics, and the general ecological impoverishment of rural environments in developed nations.

These considerations recently led to the development of a single honors degree in agroforestry at Bangor. The degree is offered jointly by the departments of agriculture and forestry, and its aim is to provide a sound, integrated education in both subjects.

The course retains a fully professional, land management bias, and offers opportunities for field courses and projects which cut across the rigid lines that have traditionally divided agriculture and forestry. In addition, the course and its concomitant research programs, draw both agriculture and forestry closer to the developing disciplines of environmental conservation and management. For example, candidates attaining the M.Sc. degree in environmental forestry can also specialize in agroforestry.

The department at Bangor has close contact with national institutions throughout the tropics, and with the international agencies such as FAO, UNEP, IBPGR, IUCN, ICRAF and the International Agricultural Research Centers [Roche 1988].

There are other institutions in Europe offering education in agroforestry. Most of them, however, restrict themselves to agroforestry definitions, basic rules and a few lectures within forestry or agricultural programs. For example, all three forestry faculties in the Federal Republic of Germany offer some agroforestry. Freiburg and Goettingen have been most active, and a remarkable number of these have emerged from postgraduate and graduate research. The University of Hamburg offers lectures in agroforestry within its program of World Forestry and provides instruction to dissertation students including those located at other universities. The number of students interested in agroforestry postgraduate studies is growing very fast, and the demand for trained staff is increasing. Thus, university training in agroforestry should be intensified, coordinated and given a higher general recognition.

Perceived needs for education

Institutions

A survey in 1984 revealed a total of 30 scientific institutions in the Federal Republic of Germany which had on-going research projects in agroforestry (the number may have increased since then). Some of the institutions contribute to training and/or education at various levels. Most research projects refer to tropical countries and are in some way linked with governmental, international or non-governmental organization (NGO) technical cooperation activities (DSE, 1987). This link is of value, because research and practical implementation/monitoring of results are closely interrelated, and good working conditions are provided. However, the various projects depend almost entirely on the "chance of the hour" and on specific personal experiences and relations. No general systematic concept exists so far, and there is no coordination or effective network outside of individual contacts.

The situation is, as far as one may generalize, very similar in other European countries.

Research

Agroforestry research concentrates on solving problems in tropical regions or on identifying practices that have been applied traditionally by rural people overseas. Very little has been done with a focus on Europe or on developing agroforestry as a discipline. An analysis of research projects in the Federal Republic of Germany (DSE, 1987) revealed that 40% of all agroforestry projects originated from forestry, 70% concentrated on sustainable management of natural resources and/or food production, 80% were field research projects, and 44% were located in the lowest among the less developed countries (LLDCs) (almost two thirds of these in Africa). In most European countries, the major agroforestry initiatives appear to originate from forestry, whereas agriculture, animal husbandry, sociology, etc., contribute to a lesser extent. An underexplored but promising area is cooperation with agro-industries.

Recommended guidelines for education and training in agroforestry

Education and training at the university level in Europe should be based on and closely related to research; this is a tradition in academia which should be maintained. Teaching would be poor if not inspired and constantly developed further through research. This is mentioned expressly because there may be some temptation to rely almost entirely on research findings from international centers or overseas institutions, which may not be very relevant.

On the other hand, research is stimulated through demands arising from teaching at all levels, and extension work. In Europe there may be specific advantages for basic research, because of existing institutional structures and available scientists. In some ways this should promote 'export' of results in basic research from Europe and 'import' of experience in applied and adaptive research from the tropics.

So far in Europe, a pronounced bias exists for laboratory research in agroforestry, whereas on-station and on-farm research are carried out overseas. However, both should be strengthened within Europe in order to meet the demands.

Many experts speak about the socio-economic aspects of agroforestry, about social forestry, etc., but very little has been done in terms of research or teaching in social issues. Finally, and this may be a specific challenge for

European research and education, the broad sector of documenting and evaluating agroforestry practices needs to be further developed.

One of the recommendations of the 1982 workshop on agroforestry education held at ICRAF [Zulberti 1987] should be taken into serious consideration in this context: 'One practical way to assist the development of agroforestry as a subject may be to encourage the 'twinning' of appropriate institutions, for example, those of forestry, agriculture, applied ecology, etc.' In education, both partners, the European and the tropical countries, could essentially gain from mutual experience in solving extremely different problems of integrated land use and in making agroforestry a modern, future-oriented option of sustainable resource management.

References

DSE (1987) Agroforstwirtschaft in den Tropen und Subtropen (Agroforestry in the tropics and subtropics). DSE/ZEL, Feldafing, F.R.G.

Roche L (1988) Letter to the editor. Agroforestry Systems 6: 179–180

Zulberti E (ed) (1987) Professional Education in Agroforestry. ICRAF, Nairobi

Agroforestry Systems **12**: 97–102, 1990.

Regional/Country report

Status of agroforestry education in India

KIRTI SINGH
Narenda Deva University of Agriculture and Technology, Faizabad, U.P., India; Present address: Himachal Pradesh Agricultural University, Palampur 176 062, India

Abstract. India is perhaps the world leader in development of agroforestry education, training and research. The Indian Council of Agricultural Research sanctioned an All-India Coordinated Research Project on agroforestry in 1983, to be headquartered in Delhi but with research centers in 20 other locations countrywide. The agricultural universities in India have a major role to play, with all institutions having agroforestry teaching programs by 1990. At this point there is a great shortage of faculty trained in agroforestry. Demand for qualified graduates with this training is only going to increase.

Introduction

The total geographical area of India is about 328.8 million hectares, out of which forests occupy 67.4 million hectares. Agroforestry as a land use system is as old as agriculture. In a country like India with varied agroclimatic zones, various combinations of trees with arable crops, fruits, and animal husbandry exist [Singh 1987; Tejwani 1987; Nair and Sreedharan, 1986].

Growing field crops like pearl millet, legumes and oil seeds in a field predominated by 'khejri' (*Prosopis cineraria*) and 'bordi' (*Zizyphus nummularia*) are examples of traditional agroforestry of Rajasthan, Punjab, Haryana and Gujarat states. Another classical example is the cultivation of large cardamom (*Amomum subulatum*) in combination with *Alnus nepalensis* in the central Himalayas. Various tree crops combinations in 'jhum', or shifting cultivation, in the North Eastern Hill Region have long been practiced. Growing tree species, such as *Dalbergia sissoo*, *Azadirachta indica*, *Acacia nilotica*, *Grewia optiva*, *Morus alba* and *Ficus* spp., on the borders of fields to meet local demands for timber, fodder and fuel is a common practice throughout the country. Homesteads of Kerala, with combinations of fruits, spices and vegetables, are unique.

Agroforestry education

Natural forest resources in India are diminishing at alarming rates, because they are unscientifically exploited to meet the needs for fuel, fodder, timber, etc. of increasing population. Decline of forests greatly influences rural poverty. In areas where the villages are located away from a green belt, human resources are mainly spent collecting fodder and fuel rather than on farming or exploitation of forest products for economic benefit.

The National Commission on Agriculture of India has estimated the consumption of fuelwood in the country to be 300 million m^3 in the year 1980, with a deficit of about 100 million m^3 [NCA 1976]. To meet this demand of 100 million m^3 of fuel wood, it is estimated that about 3 million ha per year is needed to be planted with fuelwood species over the next 10 years. The estimated yield of firewood from the forest (51 million t) and from private lands (26 million t) add up to 77 million t. This leaves a deficit of 33 million t in 1980, 44 million t in 1985 and 59 million t in the year 2000. Unless the problem is addressed adequately, alleviation of rural poverty will be nearly impossible. In this context, the prediction 'there may not be famine for food but for firewood to cook' seems to be true.

By the year 2000, with a projected population of about a billion people, a billion tons of milk would be needed in India. To meet the milk and meat requirement of this growing population more animals and increased forage production are necessary.

The increased human and animal population will cause acute shortages of fuel, fodder and food. This forces planners, scientists and educators to think seriously on various aspects of education, research and extension programs in agroforestry for the benefit of the rural masses. This is especially relevant because there is no scope of increasing the available land area.

In India, agroforestry research on scientific lines started in 1979, when a seminar was organized by the Indian Council of Agricultural Research (ICAR) at Imphal, Manipur [ICAR 1979]. The seminar, suggested the research direction and organizational pattern for strengthening agroforestry research and education. Establishment of an All-India Coordinated Research Project with centers in states and union territories for a comprehensive study on the various aspects of agroforestry was recommended. Consequently, the Indian Council of Agricultural Research sanctioned an All-India Coordinated Research Project on agroforestry. A coordinating cell in (ICAR headquarters) New Delhi, and centers at 20 places in different agroecological regions were established under the Sixth and Seventh Five Year Development Plans in the 1980's.

Later it was realized that without education in agroforestry as a land use system, it would be very difficult to solve the problem of fuel and fodder shortages and to carry out satisfactory research and extension activities. Agroforestry as a land use system is an essential need for marginal and small farmers. Agroforestry education is needed for incorporating recent technical

Table 1. Annual intake capacity of state agricultural universities offering degree programs in forestry

Year of establishment of Forestry department	No. of Agric. Universities offering degree programs in Forestry	Intake capacity		
		B.Sc.	M.Sc.	Ph.D.
1982	1	20		
1985	5	105	4	5
1986	8	147	6	—
	14	272	10	5

know-how on the subject, selection of agroforestry systems according to the specific needs of each area, interaction of different agroforestry components, selection and planting of multi-purpose trees (MPT) and nitrogen fixing trees (NFT), and selection and planting of agroforestry systems in wastelands.

Agroforestry was in practice in India long before the scientific approach to it was realized. Various indigenous agroforestry systems were included in curricula from the beginning of agricultural education in India. Many institutions still teach it to undergraduate and postgraduate students of agronomy, soil conservation and horticulture. Until 1979, the importance of this aspect was not fully realized because shortage of fuel and fodder from natural forests and pasture resources was still not obvious.

Recent developments

Presently, there are 26 agricultural universities and one horticultural university situated in different states of India. Agroforestry education was started formally with the establishment of forestry departments in different agricultural universities. The first BSc forestry program was started during 1982 at Birsa Agricultural University, Ranchi (Bihar). During 1985–86, 13 more agricultural universities became involved in the forestry teaching program. Agroforestry is only part of the forestry teaching program. The average intake of students to these programs is about 20 per year (Table 1). Kerala Agricultural University and Dr. Y.S. Parmar University of Horticulture and Forestry, Solan (Nauni), Himachal Pradesh also offer MSc degree programs in forestry, with an annual intake of four to five students. Recently, a PhD program in forestry was started at Solan (Nauni).

A three credit course in agroforestry has been included in the undergraduate forestry program of the agricultural universities (Appendix I). By the end of the Seventh Five Year Plan (1990), all the agricultural universities in India will have agroforestry teaching programs. In 1987, the Indian Council of Forestry Research and Education was established at Dehra Dun. The relationship of this council with ICAR with regard to forestry and agroforestry education and research is not yet properly defined.

At the Narendra Deva University of Agriculture and Technology (NDUAT), Faizabad, a compulsory course of two credits along with an elective course of two credits in agroforestry (Appendix II) were started as a part of the undergraduate agricultural degree program in 1978 when the college of agriculture was established. This was done to promote agroforestry in rural areas through the graduates. Teaching and practical training facilities have since been created by developing agroforestry research and instructional farms where various agroforestry models have been developed in an area of about 200 hectares.

Based on our experiences of about 10 years, the following suggestions are made for further development of agroforestry education and training in agricultural universities in India:

1. Agroforestry course should be made an integral part of the BSc programs in every agricultural university.
2. Credits in agroforestry should be increased in BSc agriculture and forestry programs.
3. Curricula should be designed in an integrated manner involving other disciplines such as horiculture, agronomy, crop physiology, soil science, agricultural economics, extension education and rural sociology.
4. Horticultural crops, fibre crops and medicinal and aromatic plants in agroforestry systems should be given more emphasis.
5. Wasteland utilization through different cropping models/agroforestry systems, depending on needs of the area, should be given more importance in education and training.
6. Agroforestry research and education in problematic soils like sodic and acidic soils need greater attention.
7. Agroforestry education needs to be promoted by governmental agencies through offering separate agroforestry scholarships and fellowships.
8. Courses and research should be designed in consideration of the suitability of agroforestry systems in specific agroclimatic regions.
9. Steps should be taken to impart short- and long-term training to those responsible for teaching agroforestry in the Agricultural Universities.
10. Summer institutes/refresher courses should be organized covering major

components of agroforestry and their interactions.
11. More subject matter specialists and extension workers involved at district, block or village level in agroforestry work must be trained.
12. Effective linkage among various organizations involved in agroforestry education in the country must be established.

References

NCA (1976) National Commission on Agriculture, India. Government of India, New Delhi
Nair MA and Sreedharan C (1986) Agroforestry farming systems in the homestead of Kerala, southern India. Agroforestry Systems 4: 339–363
Singh GB (1987) Agroforestry in the Indian subcontinent: past, present and future. In: Steppler HA and Nair PKR (eds), Agroforestry: A Decade of Development, pp 117–138. International Council for Research in Agroforestry, Nairobi
Tejwani KG (1987) Agroforestry practices and research in India. In: Gholz HL (ed). Agroforestry: Realities, Possibilities and Potentials, pp 109–136. Martinus Nijhoff Publishers, The Netherlands

Appendix 1

Forestry four-year degree courses at State Agricultural Universities

Agroforestry course in 3rd year

FOR 333 *Agroforestry* (2 + 1)

Definition and principles of agroforestry, current trend in agroforestry – Role of farm forestry in rural development – Economic gains from agroforestry systems – Social factors which influence agroforestry.

Forestry for rural communities – timber, fodder, fruit and fuel. Objectives – survey and diagnosis of agroforestry systems. Agroforestry systems – advantages and constraints.

Interaction of trees with field crops and pastures. Fodder farming, trees, grasses, legumes, nutrient content of different fodder and pasture species.

Management and protection of farm woodlots – Afforestation of village common lands – Definition and functions of windbreak and shelter belts.

Practical

Exercises that evaluate and provide experience in implementing agroforestry systems.

Appendix 2

Agroforestry courses at Narendra Deva University of Agriculture & Technology

Elective course

AF 301 *Elementary forestry* (1 + 1)

Definition and basic concept of farm forestry, silviculture, agroforestry systems, potential areas for farm forestry, forestry plants for timber, fruits, fodder, fuel, material for fencing, windbreak, etc. Industrial needs for wood, choice of forest species, general principles of nursery management and tree crop husbandry. Culture of trees and systems of their culture.

Practical

Identification of various plant materials utilized for forestry, their propagation, plantation management and control of insect pest and diseases. Practical concept for the utilization of wasteland.

AF401 *Farm forestry* (1 + 1)

Forests, their economical and ecological importance, selection of site and soil for establishing farm forest, preparation of land, planting of forest plants for the best utilization of available land. Principles and practices involved in the cultivation of *bamboo*, *Casuriana*, *Acacias*, *Eucalyptus* and other miscellaneous trees such as *shisham*, tamerind, *Neem*, etc. economics of farm forestry, *Van Mahotsav* and roadside plantation.

Practical

Practice of plant raising in nursery, digging of pits, soil treatment, plantation and management of above mentioned plants.

Agroforestry Systems **12**: 103–106, 1990.

Regional/Country report

The status of professional agroforestry education and training in Indonesia

ACHMAD SUMITRO
Forestry Department, Gadjah Mada University, Yogyakarta, Indonesia

Abstract. Although agroforestry has been practiced in Indonesia for a long time, only recently has it received the attention of professional foresters and become a subject of research and education. Several agroforestry projects and programs have recently been introduced by the Indonesian government, and trained people, especially workers, are needed in large numbers. To cope with this demand, initially foresters interested in broadening their technical knowledge were given training in agroforestry using internationally available training materials. Simultaneously, the faculties of forestry at Gadjah Mada University, Yogyakarta, and the Institute of Agriculture, Bogor started agroforestry as an elective undergraduate-level course. Currently, eight provincial forestry faculties in outer islands also teach agroforestry with special emphasis on their local needs such as management of shifting cultivation. Although no graduate-level courses are offered, many graduate students in forestry undertake research on agroforestry-related topics. A serious constraint to the development of agroforestry education in Indonesia is the language barrier, which prevents Indonesian students from having easy access to international literature.

Introduction

Agroforestry as a type of land use has been practiced in Indonesia for a long time. Well-established agricultural systems in which trees play a major role are confined to certain areas due to differences in population density, culture and environment. In the uplands of the populous islands such as Java and Bali, extensive and well-managed agroforestry systems are common now, although problems such as deforestation, erosion, siltation and flooding still exist. In the forests of Java, there is also a long history of mixed land use. As early as the 19th century the 'taungya system' was used for establishing teak plantations. On the other islands, agroforestry systems are not well-developed, and shifting cultivation is a major problem.

Consequently, although agroforestry has been well known in Indonesia, only recently has it received the attention of professional foresters, and become a subject of research and education. It is now recognized and believed that agroforestry is useful to mitigate problems and conflicts of land use in subsistence agriculture under environmentally fragile conditions. Agroforestry can help reduce soil deterioration and settle shifting cultivators on a permanent land-use basis.

However, established concepts and techniques have to be analyzed and new socially acceptable approaches introduced and implemented. Also, the scope of the taungya system in plantations has to be broadened, technically as well as socially, to cope with the more complex problems which are pressing these forests.

Several agroforestry programs and projects have recently been introduced by the Indonesian government with international (external) support. In order to undertake these programs, trained experts are needed in large numbers, especially as extension workers, because agroforestry requires active involvement with farmers. Participants of such training programs are initially recruited from foresters who are interested in broadening their technical knowledge into the more interdisciplinary-field of agroforestry.

Many research projects and seminars are also held to develop a 'new approach', influenced by scientists or other experts. For example, in 1980 a seminar on 'Some Experiences of Agroforestry on Java' (English summary only) was initiated and organized by the Faculty of Forestry, Gadjah Mada University, and the Department of Forest Management, University of Agriculture, Wageningen, the Netherlands. The seminar made a positive contribution to the enhancement of further research and better education of agroforestry in Indonesia. It reviewed the state of the art of agroforestry on Java, which has been practiced traditionally by the people, as well as several watershed development programs and teak forest plantation management.

Recognizing the experience and supported by theories and models from the International Council for Research in Agroforestry (ICRAF) and elsewhere, agroforestry in Indonesia has found a firm grip in the education system and has become a solid integrative classroom subject.

Present status of agroforestry education

With the increasing awareness of the importance of interdisciplinary views of problem solving in forestry in a changing environment, agroforestry requires at least the knowledge of agricultural cropping practices, tree growth, ecology, (micro) economics and sociology. It is accepted now as a teaching subject in many schools of forestry in Indonesia.

The faculty of forestry at Gadjah Mada University has been offering agroforestry as an elective course. It accounts for three of the total 160 required credits. It also receives ample attention in forest management, soil conservation, watershed management, sociology and economics courses. In the future, agroforestry may become a special orientation in the graduate program.

The Institute of Agriculture at Bogor offers a course in Social Forestry in the 6th semester. This course includes agroforestry in a broader sense: rural welfare, land rehabilitation, shifting cultivation, small private forest holder, rural industry, etc.

Besides those two major faculties of forestry, there are eight other provincial forestry faculties in the outer islands which follow a curriculum similar to that in Java. But since these provincial faculties are located on the islands, which still have large areas of tropical moist forest, their agroforestry teaching and research are mainly about shifting cultivation practice.

Agroforestry courses offered through the faculty of agriculture and animal husbandry do not exist yet, although similar courses, such as hill-farming-system and agro-eco-development are sometimes offered. As a matter of fact, some general agriculturalists are reluctant to accept the term agroforestry.

Even though the course of agroforestry in the formal forestry curricula is still very limited, the number of students conducting research in agroforestry is increasing. This means that university lecturers should be more open to the interdisciplinary nature of the subject. In some cases, the teaching and thesis consultation on agroforestry are conducted by joint instructors/advisors.

A graduate level forestry course in Indonesia is still in its early development. No specific courses or programs in agroforestry are offered, but many topics about agroforestry are taken for Masters or Doctoral thesis research.

International cooperation

The agroforestry theme is commonly used as an important component in rural and upland development projects in Indonesia. The program usually consists of field research, often incorporated in education programs, training, demonstration plots, publications, etc.

Currently, in East Java, a Dutch-sponsored watershed management project is conducting limited research and training in agroforestry of the upland Kalikonto watershed. Besides, the Ford Foundation, in cooperation with the Department of Forestry, is establishing several social forestry projects on Java.

In 1979, the faculty of forestry at Gadjah Mada University established a joint effort with the Department of Forest Management, University of Agriculture, Wageningen, the Netherlands. One of the activities is dealing with agroforestry study in a critical district of the province of Yogyakarta. A unique part of the undertaking is a joint effort in cooperative research between Dutch students who come to the area for three to four months work with Indonesian students in the field to collect data for their individual theses.

Possible development of agroforestry education

The demand for trained agroforesters in Indonesia has the potential to be quite high. The large dry land areas under forest and high vegetation in Indonesia are about 143 million hectares or 65% of the total land. The increasing problem of shifting cultivation in the tropical moist forest in the outer island and the increasing pressure on forested land in Java are the two key factors for the increase in the demand. The current demand for graduate level agroforesters is between 50 to 100 professionals and for technicians more than 1000.

One important constraint of agroforestry education in Indonesia is the language barrier. This inhibits communication as well as the exchange of experiences and materials internationally. Another constraint is the lack of long-term consistent interpretation of agroforestry practices and designed experiments. Political support and governmental policies aiding the development of agroforestry practices on a national scale are very promising. Although budget is a constraint, many institutes have good facilities and trained experts. This aids in the development of agroforestry education in quantity and quality in Indonesia to meet national development objectives.

Agroforestry Systems **12**: 107–114, 1990.

Regional/Country report

Agroforestry education and training in Latin America

JEAN C.L. DUBOIS
Rua Redentor 275, Apt. 401, Ipanema, 22421 Rio-de-Janeiro, RJ, Brazil

Abstract. The Center for Research and Education in Tropical Agriculture (CATIE), Costa Rica, is the premier institution in agroforestry research training and education in Latin America, and most agroforestry professionals in Latin America were, and still are, trained at CATIE. Additionally, forestry faculties in some Latin American universities have undergraduate courses in agroforestry and offer opportunities to present theses on agroforestry-related subjects for MS and PhD degrees. A 1982 survey on university-level agroforestry education in Argentina, Brazil, Chile, Colombia, and Peru showed that at that time most faculties had no organized programs in agroforestry. A 1988 updating of the information from Brazil indicated that although some universities had since introduced agroforestry courses, the majority still had no substantial interest in it. In existing agroforestry educational programs, there is room for broadening the scope of curriculum content. As regards training in agroforestry in Latin America, CATIE continues to be the foremost institution.

Introduction

The status of professional education in agroforestry in Central America has been reviewed previously [Budowski 1982, 1987; CATIE 1987]. The agroforestry education and training facilities existing at CATIE (Center for Research and Education in Tropical Agriculture, Turrialba, Costa Rica) are still the best in Latin America. Many, if not most, professors involved today in agroforestry higher-education elsewhere in Latin America and the Caribbean have attended CATIE for in-service training and/or short specialization courses. Other papers on agroforestry education in Latin America include Beer and Somarriba [1984], Fassbender [1984] and Major, Budowski and Borel [1985].

More recently, a comprehensive textbook was published in Spanish by the Organization of Tropical Studies (OTS) and CATIE [Montagnini et al. 1986]. This volume offers excellent material for education and training in agroforestry as regards Latin American conditions, with special reference to Costa Rica and some other Central American countries. CATIE has also organized several workshops, training courses and seminars during the course of the last ten years, even outside Costa Rica (e.g., Mexico, Colombia, Guatemala and Honduras), and has made the most concerted effort in Latin America in documenting agroforestry practices.

Within the framework of the International Workshop on Professional Education in Agroforestry (ICRAF/DSE) held in Nairobi, December 1982 [Zulberti, 1987], an extensive survey was carried out on university-level agroforestry education in South America. A questionnaire was sent to 131 institutions in 13 countries. Answers were received from 48 institutions (36.6% response), but only 19 institutions (14.5%) returned fully-answered questionnaires; complete responses represented five countries: Argentina (2 institutions), Brazil (10), Chile (1), Colombia (4) and Peru (2). The relatively low response indicated that at that time most faculties and/or departments of forestry and agronomy in South America had no programs or, perhaps, interest in agroforestry. Among the possible reasons are: intellectual isolation, poor local libraries (due mainly to lack of funds), staff members without adequate knowledge of foreign languages (including English, in which most information has been published), a pervasive 'routine-syndrome' (a characteristic consequence of intellectual isolation and lack of funds), and insufficient contacts with native, cabocio and small-farmer communities where traditional agroforestry production systems are still in use.

The 1982 survey results

The results of the 1982 survey are briefly analyzed below, on the basis of the information supplied by the five mentioned countries. As regards Brazil, more recent information is added, derived from a complementary survey made in 1988 with information supplied by 10 higher-education institutions, out of the 17 initially contacted.

Argentina: In 1982, of 21 institutions contacted, two fully answered questionnaires were received from the Faculty of Agrarian Sciences, National University of Cuyo, Mendoza and from the Faculty of Agronomy, University of Buenos Aires.

No agroforestry undergraduate courses per se existed in these establishments, but aspects of agroforestry were included as parts of other courses. No MS or PhD degree programs existed in agroforestry. Measures were being taken at the University of Buenos Aires to create a Forestry Department; a staff member of this department was to take a degree in agroforestry at CATIE [1983] and it was expected that, upon his return, he would lecture on agroforestry.

Brazil: In 1982, 41 institutions received the survey questionnaire: 10 answers were received (return-rate: 24.4%). In 1988, 17 institutions were contacted and reliable information was received back from 9 institutions (return-rate: 52.9%). Results of both surveys are summarized in Table 1.

Chile: Eight questionnaires were sent out in 1982 with only one fully answered questionnaire received back from the Department of Forestry, Faculty of Agronomy, Veterinary and Forestry Sciences, University of Chile, Santiago. Part of the course on "Forest Management for Arid Regions" is devoted to silvopastoral management. The technical content of that course is based principally on the many traditional silvopastoral production systems used in the arid and semi-arid regions of Chile.

Colombia: In 1982, the questionnaire was sent to 15 universities; 4 answered (Faculty of Forest Engineering, District University of Bogata; the Faculty of Forest Engineering, University of Tolima; the Faculty of Agronomy, University of Caldas and the Faculty of Agronomy of the University of Cordoba). At degree-level, no specific agroforestry courses were given, and no agroforestry aspects were treated in other disciplines except at the District University of Bogota. The University of Tolima was taking some part in agroforestry experimental work developed in Bajo-Calima (humid tropics), and a few theses on agroforestry subjects had already been submitted or were being implemented. Among the Colombian institutions that did not answer, mention should be made of the National University of Colombia, since the Medellin-based Forestry Faculty of this institution provides some agroforestry teaching.

Peru: In 1982, 15 questionnaires were distributed; two were answered (National Agarian University of La Molina, Forestry Sciences Program; National University of Lambayeque, Agronomic Sciences Program). In both Universities some agroforestry aspects are taught in other disciplines. It was also expected that a specific, but rather short course on agroforestry would be given, beginning in 1983, as part of the curriculum for MS students in forestry sciences at La Molina, Lima. As regards the universities that did not answer, it is probable that at least two are now providing some agroforestry teaching at undergraduate level (University of the Peruvian Amazon, Iquitos, and National Agrarian University of 'La Selva', Tingo Maria).

Further information from Brazil

The inquiry made in Brazil in 1988 provides some more specific preliminary

Table 1. AF Higher education in Brazil: results of surveys in 1982 and 1988

Institution	Situation in 1982	Situation in 1988
Federal Univ. of Parana, Curitiba, Section of Agrarian Sciences, Dept. of Silviculture & Management	A: some AF spects given in other disciplines B: interested in preparing MS/PhD setups viz. submission of AF thesis	A: regular AF course is given (30 hours lecturing + 30 hours) B: students can submit a thesis on AF subject
Federal Univ. of Vicosa, Minas Gerais (Forestry Dept.)	questionnaire not answered	A: a specific AF course will be given in 1989* B: students can submit AF thesis (one presented in 1987, 4 in 1989)
Faculty of Agrarian Sciences (FCAP), Belem, Para (Forestry Department)	A: a 10 hour AF course given annually. (A 20 hour specialization course is available to new forest engineers) B: none	A: longer specific AF course given annually, from first semester 1988 (30 h. lecturing + 15 h. "practicals")*. B: none
Federal Univ. of Mato Grosso, Cuiaba (Forestry Dept.)	questionnaire not answered	A: 45 hours AF given annually (incl. 30 h. "practicals") (*). B: none
Federal Rural Univ. of Rio de Janeiro (Institute of Forests)	A: AF aspects included in other disciplines, (i.e., specific methodology, results from AF research in Brazil). B: none	A: questionnaire not answered
Univ. of Brasilia (Dept. of Forestry)	questionnaire not answered	A: nothing formal: interest exists; steps made to recruit one AF professor B: being organized as part of Management Renewable Natural Resources (with possibility to submit AF Thesis)
Federal Rural Univ. Pernambuco, Recife Dept. of Agronomy	A: 5 hours teaching on AF included in "regional forestry for agronomists"	A: as in 1982, but 6 hrs. instead of 5.(*) B: none
State Univ. of Maringa, Parana (Dept. of Agronomy)	A: AF aspects introduced progressively in two preexisting disciplines	A: AF teaching included in the course of silviculture for undergraduate agronomists. B: none

Agriculture High School, Lavras, Minas Gerais	questionnaire not answered	A: interest exists to introduce AF in the curriculum, initially as an optional discipline. B: none
State Univ. of Sao Paulo, Botucatu (Faculty of Agronomic Sciences, Dept. of Forestry)	questionnaire not answered	A: the Dept. of Forestry, of recent creation, is interested to include AF in the curriculum. B: none

A: undergraduate-level, B: MS/PhD levels.
* = course content supplied by respondent.

information on the content of existing undergraduate-level agroforestry courses (see Appendix 1).

There is much room for improvement with respect to topics already treated and, on the other hand, other essential or important topics that are not adequately covered. I have assembled a list of perceived gaps in the topic coverage, which include:

— silvopastoral systems (humid tropics) and agrosilvopastoral systems (humid and drier tropics), including choice of tree and shrub species;
— methods related to economic assessment (including comparative economic assessment of alternative land-uses);
— multiple-use forest-fallow (tree-fallow) improvement and management;
— ICRAF's Diagnostic and Design approach and on-the-farm inquiries (systems selection and integration);
— agroforestry practices and production systems used by native communities (Indians) and/or traditional non-Indian (= "caboclos") silvicolous communities;
— nutrient cycling in main classes of production systems;
— improved production techniques and integrated use of organic matter;
— regional/local perennial multiple-use species;
— live fences;
— tree-planted fire-breaks;
— windbreaks;
— potential use of lesser-known (perennial) native species (e.g., melliferous spp., palms, medicinal species, antiophidic species, species producing essential oils).

According to the survey, practical (non-lecture) exercises included in existing agroforestry courses in Brazil were mainly or entirely implemented in the form of indoor exercises or activities. In general, the institutions involved had no funds to sustain field practicals related to agroforestry education.

Training courses

The role of CATIE in training has already been stressed. Most of the beneficiaries were and are professionals working in Central America, although a substantial number of beneficiaries are also from South American countries and the Caribbean, with fewer from tropical Asia and Africa.

In South America, agroforestry short (intensive) courses have been more limited in terms of frequency and total number of participants. Among courses, specific mention should be made of two of special relevance to the South American humid tropics: (1) 'Short Course on Research Methodologies for Agroforestry in the Humid Tropics', UNU/CATIE/IICA-Tropicos, Cali, Colombia, in 1983, and (2) the 'International Course on Agroforestry Research in the Amazon Region', ICRAF/INIPA/IICA/USAID, Yurimaguas, Peru, 1985.

Conclusion

So far, most undergraduate level courses in agroforestry and opportunities to present theses on agroforestry subjects for MS or PhD degrees in Latin America are provided by faculties or departments of forestry. The participation of and benefits for agronomists and other professionals (ecologists, biologists, economists, etc.) are limited. This amounts to a rather unsatisfactory situation, and more efforts should be made so that agroforestry can develop into truly effective multidisciplinary programs.

Appendix 1.

Content of AF undergraduate level courses in Brazil (situation, second semester 1988)

Content of course	Institutions[1]				
	FU.PA FO	FU.PR FO	FU.MT FO	FU.MG FO	FU.PE AG
Total length (hours) of					
lecturing	30	30	30	30	6
practicals	15	30	15	30	0
Introduction, definition	×[2]	×	×	×	×
Historical background and relative importance of AF	×	×	×	×	×
Classification of AF systems	×	×	×	×	×
Advantages and limitations of AF production systems	×	0?	0?	0?	0?
Sequential AF systems					
(a) improved tree-fallow	0?	0?	0?	0?	0?
(b) taungya	×	×	×	×	×
Non-sequential AF systems					
(a) silvoagricultural	×	×	×	0?	×
(b) silvopastoral	×	×	×	0?	×
(c) agrosilvopastoral	0?	0?	0?	0?	0?
Technological aspects (e.g., crop requirements, crop management)	×	0?	×	×	×
Competition between forestry and agricultural components	×	0?	0?	0?	0?
Ecological aspects	0?	0?	0?	×	0?

Appendix 1 *Cont.*

Content of course	Institutions[1]				
	FU.PA FO	FU.PR FO	FU.MT FO	FU.MG FO	FU.PE AG
Economic aspects	×	×	0?	×	0?
Social aspects	×	×	0?	0?	0?
Choice of tree/shrub species and species requirements	×	0?	×	×	0?
Assessment of AF systems	0?	0?	0?	0?	×
AF systems: promotion and extension	0?	0?	0?	0?	×
Special study of traditional AF systems at local/regional level	0?	0?	0?	0?	×

[1] FU.PA/FO – Federal University of Para, Belem, Dept. of Forestry.
FU.PR/FO – Federal University of Parana, Curitiba, Dept. of Forestry.
FU.MT/FO – Federal University of Mato Grosso, Cuiaba, Dept. of Forestry.
FU.MG/FO – Federal University of Vocosa, MG., Dept. of Forestry.
FU.PE/AG – Federal Rural University of Pernambuco, Recife, Dept. of Agronomy.

[2] × = topic treated.

[3] 0? = topic not specified in official content of course.

Agroforestry Systems **12**: 115–120, 1990.

Regional/Country report

Expanding opportunities for agroforestry education in the U.S. and Canadian universities

SARAH T. WARREN and WILLIAM R. BENTLEY
Winrock International Institute for Agricultural Development, Morrilton, AR 72110, USA

Abstract. A telephone survey of 39 U.S. and Canadian University forestry schools accredited with the Society of American Foresters, conducted in November 1988, showed that 25 had courses in tropical and international forestry, and 14 university faculties had been conducting research in agroforestry. Field-oriented training in agroforestry is also offered in North America, especially for voluntary and relief organization personnel. Many of the agroforestry courses have been established recently in response to desires of students with international interests and/or experience. Most of such courses are integrated programs, initiated by forestry faculty, and most of these are seminars or colloquia incorporating the experience of both faculty and students with focus on systems and issues rather than technical processes. Ideas for agroforestry course content offered by North American forestry faculty include: integration of social and technical aspects of agroforestry and social forestry, involvement of multidisciplinary teams, focus on managing marginal lands, and consideration of the importance of agroforestry systems in North America and other temperate zones.

Introduction

Agroforestry is a useful topic for learning about development in tropical nations and for rejuvenating an understanding of farm forestry in temperate North America. Agroforestry, and the related but different topic of social forestry, require the synthesis of biological and social science concepts and empirical information. These syntheses can be applied in a broad range of land-resource problems; the topic is a rich pedagogical device. Coupled with the interest in and importance of agroforestry in development, we would expect courses to be offered in most forestry and agricultural schools in North America.

A survey on agroforestry education in North America colleges and universities was presented to the Nairobi workshop on professional education in agroforestry in 1982. Surprisingly, the survey reported only one or two agroforestry courses (Mergen and Lai, 1987). Many institutions, however, offered internationally- or tropically oriented courses in which agroforestry was included as a topic but not the focus.

Table 1. Agroforestry and international/tropical forestry course offerings in 39 SAF-accredited forestry schools in North America, 1988–1989

	Agroforestry course	International/ Tropical courses	Future plans for A/F course	Faculty A/F Research	Farm Forestry	Student demand/ Demographics
Colorado State University	Y	Y	–	Y	–	I, P, D
Duke University	N	Y	Y	N	–	–
Iowa State University	N	–	N	N	–	–
Louisiana State University	Y	–	N	N	Y	–
Michigan State University	Y	Y	–	Y	–	I, P, D
Mississippi State University	N	Y	N	N	Y	D
North Carolina State University	N	Y	Y	Y	–	I, P, D
Northern Arizona University	N	–	N	N	–	–
Ohio State University	N	Y	–	Y	–	–
Oklahoma State University	N	N	N	N	–	D
Oregon State University	Y	Y	–	Y	N	D, I
Pennsylvania State University	N	N	N	N	N	–
Purdue University	N	Y	N	N	Y	–
SUNY Syracuse	N	Y	N	N	N	I, P, D
Stephen F. Austin	N	N	N	N	Y	–
Texas A&M University	Y	Y	–	–	–	I, P, D
University of Alberta	N	Y	Y	Y	–	I, D
University of Arizona	Y	Y	–	Y	–	I, P, D
University of British Columbia	N	Y	N	N	N	–
University of Cal/Berkeley	Y	Y	–	Y	N	I, P, D
University of Florida	Y	Y	–	Y	–	I, P, D
University of Georgia	N	N	N	–	–	–
University of Idaho	Y	Y	Y	–	–	I, P, D
University of Kentucky	N	N	N	N	N	–
University of Maine/Orono	N	Y	Y	–	–	P, D
University of Mass., Amherst	N	N	N	–	–	–
University of Michigan	Y	Y	–	Y	–	I, P, D
University of Minnesota	Y	Y	–	N	–	D, P, I

University of Missouri	N	–	Y	–	–	–
University of Montana	N	N	N	N	Y	–
University of Brunswick	N	N	N	N	N	–
University of Toronto	N	N	N	–	–	–
University of Washington	N	Y	–	Y	–	–
University of Wisconsin	N	Y	–	N	–	–
Utah State University	Y	Y	–	Y	–	D, I, P
Virginia Tech	N	Y	N	N	N	–
Washington State University	Y	Y	–	Y	–	I, P, D
West Virgina University	N	N	N	N	Y	D
Yale University	Y	Y	–	Y	N	I, P, D

Y = offers course(s)
N = does not offer course(s)
I = foreign students
P = returned Peace Corps Volunteers
D = Domestic students
– = not asked, or no information available

Agroforestry education in North America

For this workshop, we contacted 39 U.S. and Canadian forestry schools by telephone in November 1988. These are a sample of the schools accredited by the Society of American Foresters and offering graduate degrees in forestry (Table 1). Of the 39 schools, 14 offered at least one course focusing on agroforestry systems and techniques. Twenty-five of the 39 schools had courses in international or tropical forestry, and 14 indicated that university faculty had been conducting agroforestry research. Cornell University's School of Natural Resources, the University of Hawaii, and the University of Guelph are among non-SAF-accredited schools with active or developing agroforestry programs.

Training in agroforestry principles and techniques is also offered in North America in the U.S. Peace Corps, Cooperative for American Relief Everywhere (CARE), and other voluntary and relief organizations; the Organization for Tropical Studies (OTS) (a consortium of about 40 North American universities), and various other agencies such as Office of International Cooperation and Development (OICD) and the Forestry Support Program of the U.S. Department of Agriculture. No formal assessment was made of the agroforestry training and education offered specifically in agricultural or range and natural resource schools, although many of these may be cross-listed with forestry schools in our survey.

Why a dramatic increase in agroforestry courses since 1982?

Agroforestry courses established in the early 1980s probably were started by faculty already conducting agroforestry or agroforestry-related research. Courses established more recently may be responding to desires of the student population. Many faculty described their students as being volunteers who recently returned from international assignments where they gained hands-on experience, or as foreign students with an interest in development. Where critical mass has developed, demand for agroforestry courses is being met. Some faculty have noted a recent shift, however, toward more of the younger, less-experienced U.S. students who are interested in international and interdisciplinary topics.

Faculty with agroforestry research interests have already been mentioned as instructors. Another set comprises those faculty members with little experience in agroforestry, but who have international interests or experience (and in several cases, a domestic agricultural or range focus) and who are able to design suitable courses. Permanent positions are being created in several schools to meet the demand for agroforestry courses. Many schools began their agroforestry courses as integrated programs with other faculties; others have achieved integration of their agroforestry courses with their associated agricultural and range science faculties. The initiative appears in most cases to have come from the forestry faculties.

Most agroforestry courses are offered as seminars or colloquia incorporating the experiences of both faculty and students. The focus seems to be on issues and systems rather than technical processes. Schools emphasizing domestic agroforestry may give greater opportunity for learning field techniques and procedures. As materials and teaching methods are developed, it would be beneficial to share them. The focus seems to be highly variable, ranging from domestic to international to tropical to combinations of the three.

Guidelines for planning courses

The following ideas are offered for consideration by North American forestry faculties in planning agroforestry courses.

— Distinguish between agroforestry as a broad set of land use systems (alley cropping, intercropping, wind breaks, fuel and fodder plantations, and so forth) and as social forestry. The latter includes policy management and group process techniques and skills required to make many agroforestry schemes work, especially when land tenure is an issue and the rural poor are target beneficiaries.

— Involve agricultural specialists, range specialists, social scientists and others in the teaching process. The concepts and knowledge needed for agroforestry systems are missing from many forestry faculties. Agroforestry is a topic where the synthesis of different systems can be demonstrated in a useful and vivid form.

— Focus on problems of managing marginal lands, many of which were forest or rangelands until recently. This includes the need to rehabilitate degraded lands and to produce valuable crops for resource-limited people.

— Demonstrate the number and variety of agroforestry systems that can be observed in North America and other temperate zones as well as those in the humid, seasonal, and arid tropics.

Reference

Mergen F and Lai C (1987) Professional education in agroforestry in North America. In: Zulberti E (ed) Professional Education in Agroforestry, pp 39–55. ICRAF, Nairobi

Agroforestry Systems **12**: 121–131, 1990.

Workshop summary and synthesis

Towards a comprehensive education and training program in agroforestry

J.P. LASSOIE
Department of Natural Resources, Cornell University, Ithaca, NY 14853-3001, USA

Abstract. As a field of study, agroforestry is directly relevant to a wide variety of individuals and offers the opportunity to bring together the broad fields of forestry and agriculture to offer the scientific underpinnings for the development of a new, comprehensive and integrative land-use strategy. There are a number of constraints that must be considered as this new field develops further. One constraint is that science tends to operate as a reduction process where subjects are divided and then subdivided as more is learned about their nature and properties. However, agroforestry programs will only be successful at solving real-world problems if an integrative and comprehensive approach is adopted and maintained. Also, there can be no single education or training model that is universally applicable, because approaches to education and training in agroforestry must be country-specific depending on ecological, socioeconomic, and cultural needs. These were some of the issues addressed by the working groups at the Workshop. A summary of their recommendations highlights some key issues and possible approaches to meeting specific future needs in the areas of agroforestry education, training, and support.

Introduction

It was exciting to participate in the Second International Workshop on Professional Education and Training in Agroforestry. During three full days, over 60 professionals from around the world worked to define the future of education and training in agroforestry. The first two days were dedicated to providing background information which prepared attendees for special working sessions during the third day. Working groups on Graduate Education, Professional Training, and Resource and Information Availability and Sharing identified key issues and proposed actions related to their particular areas. Results from these working groups formed the bases for the specific recommendations provided in the next section. These proceedings, hence, report the results of a truly 'workshop' experience, one that lead to the formulation of recommendations important to the advancement of educational opportunities in agroforestry.

Providing a summary and synthesis of such a complex and detailed workshop is not an easy task. In hopes of accomplishing this assignment as concisely as possible, this paper first addresses the position that agroforestry currently occupies as a discipline by drawing on information provided in the Keynote Papers and Regional/Country Reports presented earlier in these proceedings, and on discussions that occurred during the workshop. Once this conceptual framework is established, the output from the three working groups will be summarized. A few concluding comments will follow this summary.

Agroforestry as a discipline

Developments to date

Compared to most fields of study, agroforestry is directly relevant to a wide variety of professionals. As a newly discovered area for research and application, it offers the potential to develop and apply a holistic and comprehensive land-use paradigm that is truly interdisciplinary. However, is agroforestry a philosophy or a science? Is it the most recent developmental 'fad' or an emerging applied discipline? Should agroforestry as a 'popular science' be fashioned into an academic discipline?

Agroforestry has been practiced for thousands of years by people all over the world. Many contend that through scientific discovery and examination of these practices we are experiencing the beginning of a new land-use doctrine that is akin to both agriculture and forestry. Hence, agroforestry is actually a complex field of study that includes many specific biophysical and socioeconomic disciplines.

Agroforestry currently is where the fields of agriculture and forestry were over 100 years ago in the United States and over 200 years ago in Europe. That is, although practiced worldwide, little is known about the research, teaching and training needs required to develop and apply sound land-use practices over wide geographic areas nor does an infrastructure exist to assure that such information will be forthcoming. It is no wonder that agroforestry professionals are facing a high level of uncertainty regarding the future of this field, at least as it exists in universities and development organizations.

Disciplines and fields of study evolve to gain additional knowledge and insight into observed phenomena or to address arising problem areas. The relatively new field of 'environmental and ecological toxicology' is a good example of a problem-oriented discipline. Many disciplines, however, result from an attempt to cope with the complexities of nature and humans. For example, at one time the discipline of 'naturalist' covered the entire field of basic and applied biological sciences, plus probably medicine, physics, chemistry, and astronomy! But, as humankind grew to understand and appreciate the complexities of the biological and physical worlds, increased specialization became necessary in order to cope with this complexity. In the area of applied biology, there soon were agricultural sciences associated with supporting food production systems, followed by forestry to support the production of fiber. Of course, the need to specialize within major fields became rapidly necessary leading to specialists oriented to understanding the biology and management needs of specific products (e.g., vegetables, dairy, Christmas trees, etc.) as well as those oriented to understanding basic principles supporting production (e.g., soils, climatology, ecology, etc.). Continued specialization has occurred and a discipline like forest ecology now has sub-disciplines and sub-sub-disciplines (e.g., tree ecophysiology, water relations physiology, conifer cell water relations, etc.). So, where are the scientific roots of agroforestry? Obviously, the complex fields of agriculture and forestry are logical places to start looking.

The forestry development model has its roots in Europe, especially Germany. This region required intensive land-use strategies because of its high population density and long history of human habitation. Hence, the practice of intensive forest management for fiber production evolved here, along with an educational, training, and research infrastructure that assured its long term stability. Central Europe, then, successfully exported the practice and science of forestry to the rest of the world. The availability of forestry knowledge and techniques has been readily embraced by land-use professionals in the United States as they strived to better manage the country's forest resources during the past 100 years.

Near the end of the nineteenth century, major efforts were needed to promote and institutionalize conservation in the United States and forestry schools and forest management agencies proliferated rapidly. There were needs for professional foresters as they were best equipped to address the problems associated with degraded lands in need of reforestation. Professionals in the United States adopted European forestry practices and student training programs to meet the needs of a sparsely populated, and relatively new country. Unique extraction and utilization principles emerged which became the norms for recommendations to developing countries around the world. Extraction technology, wood utilization, and plantation management became the basis for 'international forestry' foci at forestry institutions in the United States. In exporting this model to the developing world, however, it was often found to be inadequate due to its emphasis on single-use tree species, orientation to intensive monocultures of exotic species, focus on national-level priorities, and lack of consideration of local needs and social and cultural issues.

Although starting much earlier than forestry, scientific agriculture had a similar pattern of development. In the United States, a unique complex of land grant universities and agricultural experiment stations evolved to provide the research, teaching, and extension necessary to build a highly successful food production system. This intensive agriculture model was exported to the developing world through the activities of many campus-based, 'international agriculture' programs. Though helping to fuel the 'Green Revolution', this model often was found to be inadequate. In many cases, intensive agricultural practices promoted or exacerbated environmental problems such as soil erosion, ground water contamination, and air pollution. Socioeconomic problems also have arisen due to cultural conflicts, marketing restrictions, financial and resource limitations, gender issues and land tenure constraints. Probably most important was the fact that agricultural intensification programs commonly by-passed the 'poorest of the poor' – that is, the landless and those forced to farm lands very marginal for sustainable agricultural production without substantial subsidization. Attempts to address this situation have come in the form of organic agriculture, farming systems research, and low input sustainable agriculture programs. These programs have been aimed at enhancing food production and generally have neglected many environmental issues and especially the key role that woody plants can play in maintaining the ecological integrity of agricultural systems.

Agroforestry offers the opportunity to bring together the fields of forestry and agriculture in order to offer the scientific underpinnings for the development of a new, comprehensive and integrative land-use strategy. Agroforestry also offers the opportunity for developing countries to take the lead in the identification of land-use systems important to all countries as the failings of intensification are not restricted to countries with relatively low GNPs and per capita incomes. Many exciting opportunities exist. However, although many see agroforestry as a new model for integrative land use and environmental problem-solving, there are a number of potential constraints that must be considered as this new field develops further.

Constraints to further development

Science tends to operate as a reduction process where subjects are divided and then subdivided as more is learned about their nature and properties. This is the way that humankind has dealt with the complexities of nature and the geometric increases in information about the world around us. Science, therefore, whether biological, physical or social, will have inherent problems dealing with real-world problems related to land use as solutions lie at the interfaces between disciplines. Hence, the development of agroforestry teaching, research, and extension programs will only be successful if an integrative and comprehensive approach is adopted and maintained. Care must be exercised as science has evolved to be best able to deal with minutiae while finding complex, multidisciplinary problem-solving ventures difficult, if not impossible, to successfully address. In a sense, the development of an infrastructure to support the field of agroforestry will require reversing the inherent tendencies for science to continually compartmentalize and specialize.

A second potential constraint is that educators, being usually tied to research institutions, also tend to be reductionists. University-based instruction and public education are separated based on audience characteristics and the subject matter felt to be important to their needs. Within the university, subjects are partitioned based on 'time in residence'. Outside the university structure, public education becomes stratified by the perceived technical capabilities and/or intellectual sophistication of the audience. It needs to be recognized, however, that successful educational programs must target particular needs of specific audiences and that personal learning actually exists as an 'educational continuum' rather than as discrete events. This means that education is a continuous process involving a changing set of educational needs as one moves from child to young adult to professional. Therefore, youth education, resident instruction, and professional training are all interrelated, and in applied fields such as agroforestry, both practical skills as well as understanding and knowledge are important. This balance, obviously, shifts between audiences but combines to form the criteria for establishing a teaching and research program in agroforestry. In that it is more difficult to teach basic principles than basic practice (e.g., consider the ease that most learn fire control compared to thermodynamics), agroforestry educational programs have concentrated on the skills necessary for its practice. As a result, a comprehensive research foundation is not yet available to support many agroforestry practices.

A final constraint is that no single education or training model exists for agroforestry. Approaches to education and training in agroforestry must be country-specific depending on ecological, socioeconomic, and cultural needs; the characteristics and extent of human, physical, and financial resources available; and the inherent biophysical limitations to primary productivity. Therefore, local assessments of needs and capabilities must be carried out and educational and training programs tailored to such circumstances before widespread adoption of agroforestry practices will become a reality.

These were the challenges facing the working groups at the International Workshop on Professional Education and Training in Agroforestry. Participants were charged with developing an education-training network methodologies and program elements that would support an applied scientific discipline as it develops. These will only be successful and sustainable if they are integrative with respect to the biophysical, socioeconomic, and cultural factors that combine everywhere to affect land-use practices. This will require new ways of conducting education and training programs as the old ways have failed to fully meet many of the needs of humankind. This also will mean resisting the appeal of reductionism, rewarding the accomplishments of interdisciplinary scholarship, and opening the university structure to students with diverse talents and needs.

Summary of working groups

It was not an easy task to summarize the output from day-long sessions of several groups of professionals. The following comments highlight some key issues and possible approaches to meeting specific needs in the areas of education, training, and support related to further developing agroforestry as a field of study and practice. The presentatioin will not be as detailed as the subject matter deserves, but will synthesize the output from the working groups and provide a perspective from which to view the recommendations which follow in these proceedings.

Working group on graduate education in agroforestry

Key issues

Educational opportunities for students to study agroforestry at the university level have increased in recent years, but only a few comprehensive curricula are currently available. Although there is no precise estimation of need (i.e., supply versus demand), there is a general belief that programs leading to the B.S., M.S., and/or Ph.D. degrees are warranted. The exact programs and degrees required will vary because country-specific curricula will need to reflect the needs of the country and the capabilities of the institutions involved. Programs, therefore, will vary greatly between and among developed and developing countries.

Regardless of the country, however, critical mass problems (e.g., faculty, facilities and instructional materials) exist at most institutions which greatly limit their capability to develop comprehensive agroforestry programs. This includes the need to support scholarly research which increases in importance in higher degree programs. Although specialization in the agroforestry sciences becomes an integral part of post-graduate degree programs, curricula are needed for broad-based 'generalists' who can successfully function as practitioners or leaders of interdisciplinary research and/or development teams. Owing to the applied nature of agroforestry, education and research programs must be closely associated with extension programs to assure that the former continue to addrress real-world issue and needs.

Proposed actions

Agroforestry curricula must recognize the interdisciplinary nature of this field and assure that students, regardless of their level, learn basic principles and how to apply them in real-world situations. The level of research specialization, of course, will depend on the degree program, but at least two core courses are required to support any curriculum in agroforestry. First must be a technical course which deals with the science and management of agroforestry systems. This needs to be followed by a practicum where methods and approaches to interdisciplinary team problem-solving are reviewed and practiced. Such a course would be greatly strengthened if methods were actually applied to real field problems using teams composed of students from various disciplines. These core courses must be supplemented with a breadth of specialty courses (e.g., ecology, soils, statistics, sociology, etc.). The extent to which a student concentrates on one specific discipline area to support an agroforestry focus will vary with degree programs and student interests. Whether such a curriculum becomes a major or minor area of study will vary among countries. In all cases, however, an agroforestry curriculum needs to be flexible enough to attract and accommodate students from a variety of disciplines associated with the science and practice of agroforestry.

Educational activities involving teaching and researching agroforestry will need to remain decentralized. This will assure that such programs meet the specific needs of particular countries while fully utilizing the capabilities and resources of a variety of institutions and centers. The need to network and cooperate among and between academic institutions and research centers is very important as no one location can expect to amass all the resources needed to establish and maintain a comprehensive teaching/research program as well as the diverse, long term field sites needed to support it.

It is generally felt that the land grant university model developed in the United States, which links teaching, research, and extension, would be a good infrastructure within which to develop an agroforestry education program. Care needs to be exercised to assure that such a program remains committed to addressing real-world needs, which means promoting on-farm and in-country research and training activities. In order to support an applied education and research effort in agroforestry, new approaches to establishing and maintaining a funding base are required. This will require establishing higher funding priorities for agroforestry than currently exist.

Working group on professional training in agroforestry

Key issues

Agroforestry training programs have one ultimate audience – the farmer or land user – but there are a number of secondary audiences which also must be considered. Training programs also need to be designed for professional practitioners (e.g., 'extensionists') who work directly with farmers, and for policy and decision-makers who establish the guidelines within which agroforestry must ultimately function. The successful development and training of extension staff are very cost effective as they serve as 'multipliers' who in turn reach many others. The concept of 'training' often implies the teaching of skills and techniques needed to implement agroforestry schemes. Training programs also need to stress understanding of the biophysical and socioeconomic principles underlining the practices. Otherwise, given the diversity in which agroforestry must operate, improper implementation schemes will likely occur.

Proposed actions

The training model developed in the United States and embodied in the Cooperative Extension Service provides one approach to designing meaningful training programs – ones carefully structured to meet specific audiences. Application of this model starts with the identification of a particular audience and an assessment of its informational needs. A program is then developed which targets these needs and matches the capabilities and resources of the institutions involved. Once delivered the program's effectiveness must be evaluated relative to the needs originally identified. Such an evaluation forms the basis for a refinement of the program and/or the identification of additional informational needs. It is very important to involve the farmer in this iterative training process, but not just at the problem identification and/or delivery stages. Farmers and practitioners need to be involved throughout the training process as a two-way flow of information between trainers and 'students' will assure the development of programs that meet the needs of both – trainers to be educators and farmers to be sustainable agriculturalists.

Training programs must include on-farm activities as well as demonstrations at field stations in order to emphasize the real-world application of agroforestry principles. This process can and should involve an applied research component to address problems identified during training. Developing and adopting such a structure will necessitate modifying the current land grant university approach where innovations often are developed by research scientists at field stations and the results are taken to farmers by extensionists. The diagnostic and design methodology for agroforestry projects that was developed at ICRAF by Dr. John Raintree and colleagues provides an effective approach to involving farmers and professionals in a training activity that is also capable of identifying and addressing research questions.

The complexity and multidisciplinary nature of agroforestry systems will necessitate establishing team-teaching approaches. Successful farmers should themselves be the trainers of other farmers as they will carry credibility that researchers and extensionists often lack. This will mean decentralizing training programs to involve a wide variety of professionals and practitioners in as many real-world situations as possible. Effective interactions between extensionists and researchers are needed to deliver information well supported by accurate studies and to help identify informational voids needing further investigation. Along these lines, the professional status of extension workers needs to be upgraded and brought into balance with respect to time and money dedicated to research.

Working group on resources, information availability and networking in agroforestry

Key issues

Despite the fact that agroforestry is an emerging field, there is a great diversity and quantity of resources currently available for its support. There is a rapidly growing academy of agroforestry professionals dedicated to teaching, research, and/or extension activities as well as the practice of agroforestry. In recent years, there has been a remarkable increase in the amount and quality of educational and training materials relevant to agroforestry, including journals, books, proceedings, slide sets, films and video tapes. There has also been an increase in the availability of facilities and financial resources to support agroforestry activities in many parts of the world. Despite all these resources (and promises for more in the near future) agroforestry tends to be information rich and knowledge poor – that is, basic principles of agroforestry still need to be developed. The types of information needed are not only complex due to its interdisciplinary nature; they also vary greatly by audience, country, locality, and management practice. Hence, specific assessments of needs must be undertaken that will help develop basic principles which are transferable across wide stretches of variability. Certainly, no single individual, discipline, or institution can accomplish this monumental task. Equally certain, therefore, will be the need for cooperation.

Proposed actions

Given its inherent diversity and complexity, there is a need to develop and maintain a worldwide network dedicated to providing research, education, and training information related to agroforestry. This network could be as simple as ICRAF's newsletter, or as elaborate as a new professional society. Regardless of the approach, this network should facilitate the exchange of information and individuals (e.g., practicing professionals and students) between institutions, and the enhanced access to libraries and other sources of agroforestry information. Future technologies, such as teleconferencing, laser disc information storage systems and computer networks, offer many exciting possibilities for building a world community of agroforestry scholars and practitioners.

There is a need to greatly strengthen facilities committed to agroforestry education and training. More agroforestry demonstration and/or research areas are required in a wide variety of locations in order to cope with the biophysical and socioeconomic variability inherent in the world's farming systems. An enhanced research base is needed to support agroforestry practices and such research must be closely tied to extension/training activities. A funding base for agroforestry programs should be established and support increased at all levels.

Though a wealth of general information is available about agroforestry, there are still major voids in information needed to support educational programs. A new textbook is warranted that deals with the principles of agroforestry in a systems context. Such a text should be a theory-based presentation and, therefore, have worldwide application. Hence, it must be available in various languages, at least French, Spanish, and English. The agroforestry academy must develop the infrastructure to produce educational materials that meet the needs of various audiences and specific program capabilities. This means building institutional capabilities to produce such information and developing a body of professionals not only capable of generating research information, but also capable of interpreting it for nontechnical individuals such as farmers. All of this information needs to be updated often due to the rapidly increasing research data-base which is currently evolving.

Conclusions

Concerns about the constant clashes between conservation and development have gained worldwide attention as humankind has grown to realize that its food production systems are intimately linked to its natural resource base. The basic question is how can the world's resources be developed for human welfare while conserving the ecological integrity of its terrestrial and aquatic ecosystems? Conservation in the face of development is at the base of the sustainability issue currently occupying so much attention at education and development institutions worldwide.

Many believe that one viable approach to sustainable agriculture involves the promotion of agroforestry systems. Faith in this approach arises from its applied interdisciplinary base and from the fact that it has been practiced for thousands of years by agricultural societies around the world. Agroforestry, however, is still in its infancy with respect to the infrastructure needed to promote its widespread application. Those who attended the International Workshop on Professional Education and Training in Agroforestry are working to identify and develop the research, education and training programs necessary for its maturation. Whether such efforts establish agroforestry as a scientific field similar in stature to agriculture and forestry will not be known for some time. Regardless of the outcome, however, agroforestry certainly will continue to be practiced in developing countries for many more decades to come.

Acknowledgements

Thanks are due to Louise Buck and Greg Nagle for their editorial input.

Agroforestry Systems **12**: 133–139, 1990.

Recommendations of the International Workshop on Professional Education and Training in Agroforestry, University of Florida, December 1988

P.K.R. NAIR, H.L. GHOLZ and M.L. DURYEA (Editors)

Agroforestry education, training, and availability and sharing of resources and information, were the three broad topics discussed in the workshop. Following state-of-the-art reviews on each of these topics in separate plenary sessions, the issues were considered in detail by six concurrent working groups. The individual working groups' recommendations were discussed in a plenary session, and adopted with necessary modifications. The final recommendations under each topic are as follows:

Graduate education in agroforestry

Need for agroforestry education

Student interest in agroforestry education has increased due to increased awareness of ecological, socio-economic and cultural problems and opportunities in land management; a lack of confidence in established land use practices; and a feeling that current educational opportunities are too narrowly specialized.

Workshop participants perceive a need for people trained at the PhD level to serve as faculty and research scientists, at the MS level to serve as trainers, extension specialists and project managers, and at the BS level to serve as extension agents and project implementers; there seems to be a demand for larger numbers of people trained at the BS level. Further assessment of the demand and implications for program development seems warranted.

'Centers of expertise' in agroforestry education

Rec. 1. Centers of Expertise should be encouraged to develop in a variety of settings.

1.1. Existing educational centers with strong existing support and records of graduate achievements will have the best opportunities to evolve into recognized centers.

1.2. Centers in which education, research, training and information sharing can be combined should be especially encouraged.

1.3. ICRAF should be recognized and developed as an "International Repository" for agroforestry training materials, scientific documents, etc. to assist in agroforestry education world-wide. Donor support should be provided to ICRAF to strengthen its capability to perform this leading catalytic role.

1.4. Networking among institutions involved in agroforestry should be encouraged and supported.

1.5. Donor agencies are urged to identify and aid in the development of emerging centers in various regions.

Program and curriculum development

It may be appropriate to teach agroforestry at all three degree levels, including the development of: (1) a single course or an informal set of courses available on an elective basis to promote a greater awareness of agroforestry in students with other specializations, (2) a minor program to provide a specialization in agroforestry to accompany other major degree work, or (3) a major degree in agroforestry.

It should be recognized that students will be very heterogeneous in background and that their needs and expectations will be formed by a variety of factors, so that attempts to develop uniform curricula at all institutions are not advisable.

Establishment of majors in agroforestry may be premature at institutions lacking sufficient support in existing programs. It is possible, for example, that many existing degrees in forestry or agronomy, with some modifications and improvements, can facilitate the filling of jobs requiring some agroforestry expertise.

Rec. 2. At the present time, it would be more expedient and appropriate to place emphasis on the development of generalized agroforestry education programs at the MS level, with both thesis and non-thesis options.

Rec. 3. Programs should be flexible so that students can construct appropriate course sequences to fit their backgrounds and needs.

Rec. 4. In all cases (individual courses or programs), emphasis should be placed on the integration of both the bio-physical and the socio-economic sciences. One way to address this issue is for one or more of the basic courses to be team-taught.

Rec. 5. The integration of different aspects (Recommendation 4) should stress processes, concepts and theory rather than strictly site-specific information.

Rec. 6. A key feature of graduate programs should be the existence of a core of courses required of all students.

6.1. Part of the core should include a description and technical aspects of client-oriented methods of working in interdisciplinary teams (e.g., "farming systems research", "diagnosis and design").

6.2. Basic courses should include courses in (a) natural sciences (e.g., ecology, agronomy, silviculture, soil science, etc.) as applied to agroforestry; (b) social sciences, including economics; (c) agroforestry systems and technologies, stressing the scientific and technical state-of-the-art; and (d) research approaches and methodologies with emphasis on the generation of new technologies.

6.3. Case studies should be incorporated into courses as appropriate.

Rec. 7. Mechanisms for interdisciplinary interaction among students and faculty should be developed, promoted and rewarded. The absence of an interdisciplinary approach is recognized as the major stumbling block at most institutions.

Research

Rec. 8. Field research by graduate students should be in appropriate settings. Ideally, research by overseas students in developed country institutions should be carried out in students' home countries. Advantages include increased experience on relevant systems and problems, work in realistic institutional and resource conditions, and integration into on-going research programs. Potential disadvantages include inadequate supervision, inadequate facilities, personal and professional distractions (e.g., assignment to other duties).

Rec. 9. Research should focus on principles and their application to solving particular problems, rather than on technologies useful only in some specific setting.

Rec. 10. Efforts should be made on establishing long-term studies in which a sequence of students could conduct research.

Rec. 11. Research should include interdisciplinary collaboration, with supervisory committees broad in their representation.

Rec. 12. Institutional linkages ("twinning") should be pursued to provide long-term sharing of information, research sites, expertise and experiences.

Rec. 13. Analysis of existing ("traditional") systems as well as the development of new technologies should be included in possible projects.

Rec. 14. Donor agencies should provide a portion of their support for projects to graduate student research support. This would provide a high knowledge return for each dollar invested, and would also serve to train the next generation in the most appropriate setting. A corollary is that donors should begin to evaluate projects from the standpoint of whether or not long-term research and graduate training opportunities are provided for by the recipient countries/institutions. Both perspectives require a flexibility not usually present in such projects.

Networking

Rec. 15. In all aspects of graduate education, there should be sufficient "linkages" among activities such as education, training, research, and information transfer within institutions and among different institutions.

Rec. 16. Projects for joint funding of these activities should be proposed to donors, and should prove attractive from the standpoint of maintaining programs with multiple payoffs.

Rec. 17. Exchange programs for scientists should be supported and encouraged (e.g., visiting scientist/lecturer internships).

Rec. 18. Donor support for intra-country networking should be sought to facilitate the effective development of centers of agroforestry education taking most advantage of existing talent, experience, and facilities.

Rec. 19. Linkages among English-, French-, and Spanish-speaking countries, the Soviet Union and China should be universally strengthened. The extensive agroforestry experience and knowledge and information that exist in many places are not being adequately shared because of weak or non-existing linkages.

Training in agroforestry

Goals of agroforestry training

Rec. 20. The goals of agroforestry training shall be to realize maximum benefit from agroforestry technologies and specifically to:

1. Improve the well-being of farmers
2. Meet national and global needs for food, fuel, fodder, and other resources
3. Promote sustainable land productivity

Rec. 21. Training is more effective when it is targeted towards a defined group and when trainees' knowledge is accessed and utilized. Four groups of audiences shall be identified and targeted:

1. Farmers and villagers
2. Technicians and extension workers
3. Professionals and scientists
4. Politicans and senior officials

Training of farmers and villagers

Rec. 22. The curriculum should include:

1. Basic principles
2. How-to and hands-on knowledge
3. Self-help and problem-solving skills

Specific topics for the curriculum can be identified using needs assessments of farmers, national priorities, environmental needs, and new technologies.

Rec. 23. Training should be conducted using a team approach wherever possible and should involve community members. Training sessions should be scheduled at times which are appropriate to farmers' lifestyle and held at convenient locations. One-day or at times one-week field tours and demonstration plots provide good hands-on experience.

Resource materials which are appropriate for the specific culture need to be developed and tested. Materials should be adapted to gender and age and to literacy capabilities. Regional information centers should be set up to provide information in appropriate languages.

Rec. 24. Considering that trainers are trained farmers, local extension agents, and professional, it is recommended that training programs be adapted to where the audience can go and usually these are the following:

1. On-farm sites
2. Village training centers
3. Demonstration sites

However, the establishment of internships on other sites such as farms and local research centers should be strengthened.

Training of technicians and extension workers

Technicians are defined as those, who usually do not hold college-level degrees, but are employed primarily by agencies or NGOs which have training type interactions with farmers. Extension workers are included in this category. A function of technicians is to listen to and teach farmers and to provide feedback to researchers and scientists on the needs of farmers.

Rec. 25. The curriculum should include:

1. Extension methods including needs assessment techniques, team-teaching methods, problem analysis and problem-solving skills, and monitoring and evaluation
2. Bio-physical principles such as tree/crop interactions and above- and below-ground processes
3. Socio-economic principles such as land/tree tenure, marketing, and cultural biases
4. How to find resource materials and people including interaction with researchers and how to use research results and how to establish contacts

Rec. 26. Training should use a multi-disciplinary team approach and include practical, hands-on experience accompanying classroom work. Training should be a continuing process; follow-up courses and/or on-going instruction to learn new techniques and information help reinforce information and complete the training experience.

Resource materials should include manuals and texts, slides, reference guides, case studies with data sets, and demonstration plots. It is also important to supply resource materials which can be distributed and used to teach farmers.

Rec. 27. There should be exposure to a variety of instructors. A training team might include agroforestry graduates and university faculty to teach the principles and voluntary workers and farmers to help with the applied aspects. Again, as with the training of farmers, trainees' knowledge should be accessed and utilized in training activities.

Agroforestry centers should be established to provide training materials and course announcements. Some training activities can also be accomplished at these centers and trainees may benefit from training which is done out-of-country. Applied agroforestry is best taught in the home country in the locality in which the extension work will be done.

Training of professionals and scientists

Professionals are defined as persons with at least a B.S. degree who are working as resource managers or research scientists.

Rec. 28. The objective of training in this group is to develop an appreciation and understanding of agroforestry among professionals so they may become better resource managers, and among scientists so they may more clearly perceive research needs. Recommended topics in the curriculum include:

1. Extension principles including the professional's role in program delivery
2. Bio-physical principles
3. Socio-economic principles
4. Inter-disciplinary nature of agroforestry
5. Public policy

Rec. 29. The training methods should include short courses, workshops, and observational field tours. Resource materials include manuals, texts, research publications, slides, and demonstration plots.

Rec. 30. Basic principles should be taught by scientists and professional while applications should be taught by experienced field practitioners. For research training, trainers should also be active in research and likewise for extension training, trainers should be active in extension. Training teams should include full-time agroforesters.

Rec. 31. The locations for training should be at both regional and out-of country agroforestry centers for basic topics and field locations for applied topics.

Training of politicians and senior officials

It is essential to reach this important group to enlist their support for agroforestry and to provide information for policy-making and/or influence policy changes. The need for donor agencies to enhance funding for training of this important group of people is also recognized.

Rec. 32. It is recommended that high level briefings by recognized international leaders be arranged, video supplements for a concise overview of agroforestry and training programs be organized, and tours and field demonstrations to illustrate the problems and possible solutions be undertaken.

Funding

Rec. 33. Funding for training should be an integral part of every research and development project in agroforestry.

Rec. 34. To ensure continuity of training efforts, regional training centers and accompanying materials should be developed and funded.

Rec. 35. To ensure more home country involvement and commitment to training, and to encourage continuity of effort, the donor agencies should fund home country training, either financially or in-kind.

Rec. 36. Training budgets should include funds for participants to implement some of the learned skills and activities, and for evaluation and follow-up.

Resource and information availability and sharing

Resources were defined as people, materials, and facilities. The focus in the session was on materials, specifically written information including primary and grey literature, case studies, data sets, and bibliographies. Two top priority activities are proposed, and additional information needs and methods of sharing information are suggested.

Rec. 37. A worldwide agroforestry network should be initiated as a priority activity to enhance agroforestry education through information exchange. Stable funding must be solicited to support and continue these activities. The primary clientele will be agroforestry educators who need up-to-date information on agroforestry.

Rec. 38. It is recommended that it should be a priority activity to produce an introductory-level agroforestry textbook focusing on principles with global application, and to make it available in English, French, and Spanish. The textbook will be developed by compiling references currently used by agroforestry instructors and using a constructed instructor's guide to formulate a plan for the textbook. The Forestry Support Program of U.S. Department of Agriculture will solicit interested universities and regional instructional institutions throughout the world and make recommendations to conference participants concerning publication.

Rec. 39. Five important information needs were identified:

39.1. A qualitative inventory of materials known to be useful in agroforestry education.

39.2. An introductory textbook on the principles of agroforestry applicable to a variety of settings.

39.3. Recognition and sharing of techniques appropriate for teaching agroforestry in various settings including incorporation of demonstrations and practicals.

39.4. Timely, widely distributed notice of visiting professor and employment opportunities.

39.5. Notices of course offerings.

Rec. 40. Six critical methods for sharing information among agroforestry educators were identified:

40.1. Meetings and training sessions for agroforestry instructors.

40.2. Agroforestry education newsletter.

40.3. A professional organization of agroforestry educators.

40.4. Teaching notes column in ICRAF Magazine.

40.5. Periodic issues of *Agroforestry Systems* devoted to education.

40.6. An interactive global network for education and training in agroforestry, operating in English, French, and Spanish, including regional networks.

Agroforestry Systems **12**: 141–148, 1990.

List of participants

International Workshop on Professional Education and Training in Agroforestry, 5–8 December 1988

Dr. **J.R.W. Aluma**, Head
Department of Forestry, Makerere University, P.O. Box 7062, Kampala
Uganda

Professor **E.O. Asare**, Director
Institute of Renewable Resources University of Science & Technology
Private Mail Bag, Kumasi
Ghana

Mr. **Herbert Attaway**, Forestry Educational Consultant
Route 14, Box 554, Lake City, FL 32055
USA
Tel. (904) 752–1983

Dr. **M. Avila**, Senior Scientist
Collaborative Programs, ICRAF, P.O. Box 30677, Nairobi
Kenya

Dr. **Kamis Awang**, Dean
Faculty of Forestry, Universiti Pertanian Malaysia, UPM 43400, Serdang, Selangor, Malaysia

Dr. **Dwight Baker**
School of Forestry & Environmental Studies, Yale University, 370 Prospect St.
New Haven, CT 06511
USA
Tel (203) 432-5143

Dr. **Carlos Linares Benismon**, Professor
Silviculture, La Molina Agricultural University, Chinchon 858, Letra A, San Isidro, Aptdo. 18-1393 Lima
Peru
Tel. 51-14-352035

Dr. **Arnim Bonnemann**
Section: Forestry and Agroforestry Production, Project: GTZ-Agroforestry Cooperation with CATIE, 7170 CATIE, Aptdo. A26, Turrialba
Costa Rica
Telex: 8005 CATIE CR

Dr. **K. Buhr**
Agronomy Department, 304 Newell Hall, University of Florida, Gainesville, FL 32611-0303
USA
Tel. (904) 392-1823

Dr. **Charles Davey**
Department of Forestry, North Carolina State University, Raleigh, NC 27695-8002
USA

Dr. **Jean C.L. Dubois**
Rua Redentor 275, Apt. 401, Ipanema, 22421 Rio-de-Janeiro RJ
Brazil
Tel.: (021) 5110863 (off)
(021) 5110863 (home)

Dr. **M.L. Duryea**
Department of Forestry, University of Florida, 118 Newins Ziegler Hall, Gainesville, FL 32611-0303
USA
Tel. (904) 392-5420

Dr. **K.C. Ewel**
Department of Forestry, University of Florida, 118 Newins-Ziegler Hall, Gainesville, FL 32611-0303
USA
Tel. (904) 392-4851

Dr. **P. Ffolliott**, Professor
Dryland Forestry Program, School of Renewable Natural Resources, College of Agriculture 325 Biology Sciences East Building, University of Arizona, Tucson, Arizona 85721
USA
Tel. (062) 612-7276

Dr. **R.F. Fisher**, Head
Department of Forestry, College of Natural Resources, Utah St. University, Logan, Utah 84322-5215
USA
Tel. (801) 750-2455/750-2456

Dr. **JoEllen Force**, Associate Professor
Department of Forest Resources, College of Forestry, Wildlife & Range Sciences, University of Idaho, Moscow, ID 83843
USA
Tel. (208) 885-7311

Dr. **J. Fownes**
Department of Agronomy & Soils, College of Tropical Agriculture & Human Resources 1910 East-West Road, Honolulu, HI 96822
USA
Tel. (808) 948-7508/948-8708

Mr. **Don Gasser**
Department of Forest & Resource Management, University of California, 145 Mulford Hall, Berkeley, CA 94720
USA
Tel. (415) 642-5059

Dr. **Thomas Geary**, Training and Education Coordinator
Forestry Support Program for USAID, USDA Forest Service, P.O. Box 96090 Washington, D.C. 20090-6090
USA
Tel. (703) 235-2432

Dr. **H. Gholz**
Department of Forestry, 118 Newins-Ziegler Hall, University of Florida, Gainesville, Fl 32611-0303
USA
Tel. (904) 392-4851

Dr. **M. Gold**, Director
International Forestry Program, Department of Forestry, 126 Natural Resources Building, East Lansing, Mich. 48824-1222
USA
Tel. (517) 355-0090

Dr. **K. Gopikumar**, Associate Professor
College of Forestry, Vellanikkara-Trichur, Kerala, 680654
India

Dr. **W.P. Groeneveld**
Institute for Anthropological Environmental Studies, C.P. 791, 78.900 Porto Velho, Rondonia
Brazil
Tel. (069) 2223-1560

Mr. **Olle Gustafsson**
International Rural Development Centre, Swedish University of Agricultural Sciences, P.O. Box 7005, S-750 07 Uppsala
Sweden
Telephone: 18-17-10 00
Telex: 12442 FOTEX S

Dr. **Linda Hardesty**
Department of Natural Resource Sciences, Washington State University, Pullman, WA 99164-6410
USA
Tel. (509) 335-6166

Dr. **Joerg Henninger**
Mision Forestal Alemana, Universidad Nacional de Asuncion Casilla de Correos 471 Asuncion
Paraguay
Tel. 501 516-022-3189

Dr. **P. Hildebrand**
Food and Resource Economics Department, 1157 McCarty Hall, University of Florida Gainesville FL 32611-0303
USA
Tel. (904) 392-1854

Dr. **J. Hoffman**
USDA/OICD/ITD, R. 258 McG, Washington, D.C. 20250-4300
USA
Tel. (202) 653-7777

Dr. **D. Hubbell**
Soil Science Department, 106 Newell Hall, University of Florida, Gainesville FL 32611-0303
USA
Tel. (904) 392-1951

Dr. **D.V. Johnson**, Agroforestry Coordinator
Forestry Support Program, (USDA Forest Service), P.O. Box 96090, Washington, D.C. 20090
USA
Tel. (703) 235-2432

Dr. **B.T. Kang**, Soil Scientist
Resource and Crop Management Program, International Institute of Tropical Agriculture, Ibadan
Nigeria
c/o L.W. Lambourn & Co., Carolyn House, 26 Dingwall Road, Croydon CR9 3EE
England
Telex: TDS IBA NG 2031 (Box 015) or
TROPIB NG 31417

Dr. **Donald Kass**, Interim Agroforestry Coordinator Program in Sustained Agricultural Production and Development
CATIE, Turrialba
Costa Rica

Dr. **B. Khangia**, Associate Professor
Department of Forestry, Biswanath Agricultural College Assam Agricultural University, Biswanath Chariali – 784176, Assam
India

Dr. **S. Kumar**
College of Forestry, University of Horticulture and Forestry, P.O. Box Nauni Solan (H.P.) 173230
India

Dr. **James P. Lassoie**, Chairman
Department of Natural Resources, Fernow Hall, Cornell University
College of Agriculture & Life Sciences, Ithaca, N.Y. 14853-3001
USA
Tel. (607) 255-2298

Mr. **Ken MacDicken**
F/FRED Coordinating Unit, P.O. Box 1038, Kasetstart Post Office, Bangkok 10903
Thailand

Dr. **Arnett C. Mace, Jr.**, Director
School of Forest Resources and Conservation, University of Florida, 118 Newins Ziegler Hall, Gainesville, FL 32611-0303
USA
Tel. (904) 392-1791

Mr. **William Macklin**
Nitrogen Fixing Tree Association, P.O. Box 680, Waimanalo, Hawaii 96795
USA
Tel. (808) 259-8555/259-8685
Cable NFTAHAWAII, Hawaii
Telex: 5101004385

Dr. **H.-J. von Maydell**
Institute for World Forestry and Ecology, Leuschnerstrasse 91, 2050 Hamburg 80 Postfach 80 02 10
West Germany
Tel. (9-011)-49-40-739-62419

Dr. **Peter J. Murphy**, Associate Dean (Forestry)
Faculty of Agriculture and Forestry, University of Alberta, Edmonton, Alberta T6G2P5
Canada
Tel. (403) 432-4931

Dr. **P.K.R. Nair**
Department of Forestry, University of Florida, 118 Newins-Ziegler Hall, Gainesville, Fl 32611-0303
USA
Tel. (904) 392-4851
Fax (904) 392-1707

Dr. **B.A. Naslund**
Swedish University of Agricultural Sciences, Department of Silviculture, S-90183 Umea
Sweden
Tel. 090/165856

Dr. **Hirosi Noda**
INPA, P.O. Box 478, 69000 Manaus, Amazonas
Brazil
Tel. (092) 236-9733

Dr. **Raphael E. Ole-Meiludie**, Dean
Faculty of Forestry, Sokoine University Agriculture, P.O. Box 3009, Morogoro
Tanzania

Dr. **Marcos Pena-Franjul**
University Nacional Pedro Henriquez Urena, Apartado Postal 842-2, Santo Domingo D.N.
Dominican Republic

Dr. V. **Rajagopalan**, Vice-Chancellor
Tamil Nadu Agricultural University, Coimbatore 541003
India
Tel. (91) (422) 41788
Telex: 855-360 TNAU IN

Dr. **C.P.P. Reid**, Chairman and Professor
Department of Forestry, University of Florida, 118 Newins-Ziegler Hall, Gainesville, Fl 32611-0303
USA
Tel. (904) 392-1850

Dr. **Douglas P. Richards**, Head
Department of Forestry, Mississippi State University, P.O. Drawer FR, Mississippi State, MS 39762
USA
Tel. (601) 325-2946

Dr. **Dan Rugabira**
Department of Forestry, Ministry of Agriculture, Kigali
Rwanda

Dr. **B. Runyinya**
Doyen, Faculte d'Agronomie, Universite Natural du Rwanda, Butare
Rwanda

Dr. **Stephen R. Ruth**, Professor
Department of Decision Sciences, 4400 Univ. Drive, Fairfax, Virginia 22030
USA
Tel. (703) 323-2738

Dr. **Roger Sands**, Head
School of Forestry, The University of Melbourne, Creswick, Victoria 3363
Australia
Tel. 053/61 3/452405
Telex: AA 35185 UNIMEL
Cable UNIMELB

Dr. **Marianne Schmink**
Center for Latin American Studies, 319 Grinter Hall, University of Florida, Gainesville, Fl 32611-0303
USA
Tel. (904) 392-0375

Dr. **Wang Shiji**, Director
Research Institute of Forestry, Chinese Academy of Forestry, Wan Shou Shan, Beijing
P.R. China
Tel. 281431

Dr. **Kirti Singh**, Vice-Chancellor
Narendra Deva University of Agriculture and Technology, Narendra Nagar, P.O. Kumarganj
Faizabad 224229, U.P.
India
Cable: Agriversity

Dr. **Jose Di Stefano**
Escuela de Biologia, Universidad de Costa Rica, San Jose
Costa Rica
Tel. (506) 531021 (h)
24-5616 (w)

Professor **Howard A. Steppler**, Professor Emeritus
Department of Plant Science, McDonald College, 21,111 Lakeshore Road, Ste Anne de Bellevue
PQ
Canada H9X1CO
Tel. (514) 398-7851 (w)
(514) 398-7752 (h)
Telex: 05821788

Dr. **Achmad Sumitro**, Dean
Faculty of Forestry, Gadjah Mada University, Bulaksumur, Yogyakarta
Indonesia
Telex: 25135 UGMYK IA

Mr. **Virgilio M. Viana**
Department Ciencias Florestais – ESALQ, University Sao Paulo, Piracicaba, SP. 13400
Brazil

Ms. **Sarah Warren**
Renewable Resources Management, Winrock International, Route 3, Morrilton, AR 72110
USA
Tel. (501) 727-5435

Mr. **Peter A. Williams**
University of Guelph, Department of Environmental Biology, Ontario N1G 2W1
Canada
(519) 824-4120 x-3488

Dr. **George Wilson**, Director of Research
Agricultural Development Foundation, 19 Dominica Drive, Kingston 5 Jamaica
West Indies

Dr. **W.J. Wiltbank**
Fruit Crops Department, 1137 Fifield Building, University of Florida, Gainesville, FL 32611-0303
USA
Tel. (904) 392-0071

148

Dr. **Ester Zulberti**, Principal Training Officer
ICRAF, P.O. Box 30677, Nairobi
Kenya
Tel. 254-2-521450
Telex: 22048